Contents

Dedicated to Dr. John Reagor
By coauthors, coworkers, and friends

Dr. John Reagor has given 41 years of service to Texas A&M University, the veterinary profession, and the citizens of Texas and the world. He is one of those rare individuals who is comfortable and competent whether he's in a chemistry or pathology lab, speaking at a county producer meeting, or just taking time to visit on the phone with someone who has a problem. He has amassed a lifetime of knowledge through his unique activities in research, clinical diagnostics and public service.

John: We are privileged to know and work with you. You are a scholar, a mentor and a public servant without peer.

Toxic Plants of Texas

Integrated Management Strategies to Prevent Livestock Losses

Charles R. Hart
Associate Professor and Extension Range Specialist
Texas Agricultural Extension Service

Tam Garland
Research Veterinarian
College of Veterinary Medicine
Texas A&M University

A. Catherine Barr
Assistant Toxicologist
Texas Veterinary Medical Diagnostic Lab

Bruce B. Carpenter
Associate Professor and Extension Livestock Specialist
Texas Agricultural Extension Service

John C. Reagor
Head, Diagnostic Toxicology
Texas Veterinary Medical Diagnostic Lab

Introduction

Toxic plant poisonings cause an estimated $50 million to $100 million in livestock losses in Texas every year. Losses in 1987 from one toxic plant alone—broom snakeweed—were estimated to be more than $30 million. These estimates do not include the decreased production from animals that become ill but do not die, nor do they factor in the reduced carrying capacity from loss of grazing when infested pastures cannot be used.

A key to controlling these plants is being able to identify them and to know their growth habits and toxicology, the clinical signs they cause, the methods for treating affected animals and strategies to prevent poisonings. This handbook was written to give livestock producers the information they need to reduce losses from the most common toxic plants in Texas.

Livestock producers should remember that many factors other than toxic plants can cause livestock to become ill or die. The first challenge when confronted with sick or dead animals is to determine the cause—be it related to disease, weather, nutritional status or many other possibilities and/or interactions. Those who know about toxic plants will find it easier to pinpoint the potential causes of sickness or death in livestock.

Toxic Plants of Texas includes descriptions of the 106 most common toxic plants in the state; explanations of what makes each plant toxic and how much an animal must consume to be affected; lists of animal poisoning signs; strategies for treating the animal (if known), preventing poisonings and controlling the plants; and photographs of distinguishing plant features.

The book also offers a brief overview of integrated toxic plant management, which is an approach to controlling toxic plants efficiently, economically and with the least harm to the environment.

Also included are descriptions of animal clinical signs usually associated with livestock-poisoning plants, a glossary and an index to common and scientific plant names. For quick diagnostic work, a field key at the end of the handbook cross-references animal poisoning signs with plant species.

1

The authors wish to acknowledge and thank Dr. Bill Fox and Dr. Barron Rector for assistance with plant identification, taxonomy and manuscript review. This material is based on work supported by the Cooperative State Research, Education, and Extension Service, U.S. Department of Agriculture, under agreement No. 99-EPMP-1-0655

The photographs are courtesy of Charles R. Hart, John Reagor, Tam Garland, Catherine Barr, Kelly Allred, Intermountain Herbarium, Utah State University, and USDA/ARS Poisonous Plant Research Laboratory.

IPM: Using a Variety of Methods to Control Toxic Plants

Integrated Pest Management (IPM) is a balanced, tactical approach to pest control. IPM combines several control practices, including biological, cultural and mechanical methods; monitoring of pest populations; and careful use of pesticides.

IPM involves taking action to prevent pest outbreaks when possible and to prevent excessive damage caused by pests. People who use IPM seek to control pests economically, effectively and with the least risk to the environment.

Integrated Toxic Plant Management (ITPM) applies IPM using a four-step process for managing toxic plant problems:

- Know the plants that may poison your livestock.
- Understand the clinical signs of poisoning in livestock.
- Recognize problem areas.
- Devise a management strategy for minimizing livestock losses.

Know the plants that may poison your livestock

Knowing which plants can be toxic is the first step in preventing poisonings. Photographs and descriptions, such as those in this handbook, may be helpful references. Some toxic plants are not present every year, and so do not pose annual threats.

Most poisonous plants kill animals only if they are eaten in relatively large amounts over a short period. Therefore, the dose usually determines whether poisoning occurs. Because livestock normally eat a variety of plants, rangelands in good condition provide a variety of forages and generally are safer than those in poor condition.

Other factors that may affect a plant's toxicity include:

- **Plant growth phase.** The growth phase of some plants can affect their toxicity. For example, the leaves of African rue and broad-leaf milkweed are most toxic when they are young; the reverse is true for kochia and smallhead sneezeweed.

- **Plant part.** Certain parts of some toxic plants are more toxic than the rest of the plant. For example, coyotillo seeds are much more toxic than the leaves, while all parts of lechuguilla are toxic at all stages of growth.

- **Livestock species and physiological state.** Some toxic plants are not equally toxic to all species of livestock. For example, western bitterweed is toxic to sheep but usually not to cattle or goats. Other plants may be more toxic to animals in certain physiological states. Broom snakeweed can cause abortion in pregnant livestock, and rayless goldenrod can poison suckling calves and lambs via the milk, even though the dams show no signs of poisoning.

- **Season of the year and environmental conditions.** During drought, many of the more palatable range plants wither or are unavailable, while many toxic plants remain green longer and/or respond to moisture sooner. As a result, these toxic plants may be more often consumed by livestock. Extreme environmental conditions can make some plants toxic that otherwise would not be, or increase the toxicity of other plants. Plants causing prussic acid (cyanide or HCN) poisoning, such as wild plum, mountain mahogany, and johnsongrass, remain nontoxic except during environmental extremes. Western bitterweed becomes more toxic during drought.

- **Type of soil in which the plant grows.** Broom snakeweed is toxic on sandy soils but relatively nontoxic on heavy soils. When growing in soils high in nitrogen, plants such as kochia or carelessweed may cause nitrate poisoning.

Understand the clinical signs of poisoning in livestock

The clinical signs of poisoning for each plant are included in the section on plant descriptions and management strategies. Livestock managers who are in the habit of observing animals methodically and frequently are often able to minimize plant-poisoning problems in individuals or across the herd or flock. Observe animals for behavioral changes, body condition and bodily functions.

- **Abnormal behaviors:** These may be subtle, as in isolation from the herd or flock, reduced grazing time, or other behaviors. Or they may be more obvious, as in aggressiveness or nervousness.

- **Body condition:** Pay particular attention to flesh and fat cover over key areas of the body, such as the ribs, spine, shoulder blades, and hook and pin bones.
- **Bodily functions:** Simple posture can indicate pain or discomfort. Observe ruminants for normal cud chewing while at rest. There may be problems with walking or other motor skills. Observe respiration rate and if possible, heart rate. Examine the animal's eyes for pupil function or abnormal color. Check the skin for color and know how to spot edema or swelling. Look for abnormal discharges from the mouth, nose, or eyes. If animals are sick, try to observe urine and feces for content and consistency. Compare pregnancy check rates with calf crops.

Recognizing the clinical signs associated with each plant is important in determining whether problems observed in livestock are caused by the plant. Most (if not all) native pastures contain at least one toxic plant species, but most livestock illnesses are not caused by ingestion of these plants. Poisioning should be suspected only when the animal's clinical signs are similar to those caused by the plants known to be present in the pasture.

Recognize problem areas early

Managers who recognize toxic plants and understand their toxicology can spot potentially problematic areas before they cause harm. Many plants such as smallhead sneezeweed grow in a rosette form during the winter. The number of rosettes present in the winter can indicate the severity of infestations for the next spring.

Devise a management strategy for minimizing livestock losses

Producers who recognize problem areas must devise management strategies to reduce livestock losses. Such strategies include using good range and livestock management practices, recognizing when plants need to be controlled, knowing what kind of control is best and planning for follow-up management to reduce reinfestation by targeted plants.

Good grazing management practices

These grazing management practices can help reduce or avoid livestock losses:

- In areas infested with toxic plants, decrease grazing pressure by reducing the number of animals per area of land. This allows the animals to graze selectively around toxic plants.

- During extreme environmental conditions, avoid grazing pastures infested with toxic plants. Excessively rainy periods or small rain showers during drought can cause toxic weeds to sprout rapidly and become abnormally abundant. Many toxic plants resist drought better than desirable forage species, and may be the only source of "green" forage available for some time.

- Avoid grazing pastures infested with toxic plants until enough forage from desirable plants is available. Several toxic plants are often among the first to "green-up" in spring and may remain green longer in the fall. Use these pastures during the time of year when other forage is available.

- Use the proper stocking rate for the pasture. If preferred forages are overgrazed, livestock will seek out remaining forage, including toxic plants. This is of particular concern when using high-density grazing systems.

- Maintain a few toxic-plant-free pastures to be used when other pastures are unsafe. Control methods may be needed to create such pastures, but often they already exist and the only need is to redesign fencelines or grazing rotations.

Use good livestock management practices

Suggestions for good livestock management include:

- Never keep, release, drive or bed hungry animals in or through areas known to have toxic plants. Hungry animals are much less selective of forages and can eat many toxic plants quickly.

- Graze infested areas with the type of livestock least affected by the plants present.

- Feed enough protein, energy, minerals or vitamins when needed. Deficiencies may cause livestock to eat toxic plants they normally avoid. Texas rangelands are typically deficient in phosphorus; deficiencies in phosphorus or vitamin A may cause abnormal appetites and increase consumption of toxic plants. In general, supplemental feeding is most important in

6

winter and early spring when forage is poorer.

- When possible, use animals native to the area being grazed. New livestock coming in from other areas graze less selectively until they "learn" the new vegetation. Nonnative animals are more likely to be poisoned than native animals.

Select the proper control method when needed

Control measures may be required when toxic plants become a dominant part of a pasture's vegetation. If toxicity problems occur only seasonally, it may be advantageous to concentrate on controlling plants in a few pastures. Follow any treatment with proper grazing management strategies.

These control options can be used individually or in combination:

- **Prescribed burning** does little damage to herbaceous perennial plants and usually does not control toxic plants well. However, fire may be useful where annual toxic plants occur in small areas. For example, cocklebur may be burned out of low areas like dry stock tanks. In these cases, artificial fires such as those from propane burners may be needed to burn down the vegetation. Fire also can suppress woody toxic plants such as mesquite.

- **Mechanical control methods** include chaining, railing, root plowing or grubbing. They may control some woody toxic plants, including whitebrush, oak and mesquite, but some herbaceous toxic plants, such as western bitterweed and twinleaf senna, proliferate after soil is disturbed. Where the terrain allows, some herbaceous, annual toxic plants such as smallhead sneezeweed or coulter conyza can be controlled by shredding or mowing them.

- **Biological control methods** use living organisms to feed on toxic plants. This method is particularly effective when the plant is toxic to one livestock species and not to another, because the unaffected species can be used to control the plant. Many insects are also being evaluated, and some are available for controlling a few plants. Perennial broomweed has been shown to have numerous insect enemies, as do locoweeds, but these insects are currently unavailable for commercial use.

- **Chemical control methods** are often the most economical and effective. Depending on their growth habit and susceptibility to herbicides, toxic plants may be controlled with individual plant treatments, ground or aerial broadcast application, or a combination of methods. Treating plants individually can reduce the amount of herbicide applied per acre.

When managing brush or weeds, it is important to apply the principles of integrated brush management systems, in which brush or weeds are managed with a long-term perspective. Plan for follow-up strategies to enhance initial treatments.

Note: Livestock may be much more likely to eat plants as they begin to wilt after having been treated with herbicides. Some plants become more toxic or more palatable after herbicide treatment. Do not graze livestock on treated pastures until the affected plants have completely dried. For herbicides and rates for controlling specific plants, see B-1466, *Chemical Weed and Brush Control Suggestions for Rangeland,* available from the Texas Agricultural Extension Service.

Animal Conditions

Birth defects: The occurrence of malformed offspring can be caused by a variety of factors, including the mother's consumption of plant toxins during pregnancy. Clincial signs of poisoning may go unnoticed in the dam. Plants that can cause birth defects include:

- Locoweeds, *Astragalus* and *Oxytropis* species
- Poison hemlock, *Conium maculatum*
- Singletary pea, *Lathyrus hirsutus*
- Tree tobacco, *Nicotiana glauca*
- Desert tobacco, *Nicotiana trigonophylla*

Bloat: An excessive accumulation of gases in the reticulum and rumen characterized by distension in the animal's left abdominal area. Bloat can be caused by toxic plants as well as by green, rapidly growing, nontoxic plants high in protein.

Ergot poisoning: Poisoning from ingesting ergot alkaloids. Ergot toxins are produced by parasitic fungi (*Claviceps* spp.) infesting the seed heads of certain grasses when conditions are favorable. Fungal growth replaces much of the grass seed, which turns an uncharacteristic color, and sometimes ends up as much as four times normal size. Dallisgrass, bahiagrass, small grains like ryegrass, and occasionally range grasses such as tobosa may become infected.

Two types of ergot poisoning have been described: nervous ergotism and gangrenous ergotism. Clinical signs of nervous ergotism include extreme nervous behavior and elevated body temperature, respiration rate and pulse rate. Animals exhibit a "sawhorse stance" and lean on stationary objects (or each other) for balance. Eventually they go down and are too incoordinated to rise. Gangrenous ergotism may include abortion or lameness and sloughing of the ear tips, hooves or tail swtiches.

Animals with gangrenous ergotism often must be destroyed for humane reasons.

Extremes of temperature are problematic with gangrenous ergotism. Extreme cold accentuates the constriction of blood vessels caused by the ergot and can hasten the loss of appendages. In extreme heat, the ergot-induced vasoconstriction prevents the animal from dissipating body heat, and heat intolerance becomes a potentially fatal problem.

Ketosis: A condition caused by a depletion of energy (glucose) reserves in the animal, which elevates levels of ketone bodies (metabolic byproducts of fat) in body fluids. The breath, milk or urine smells like acetone (nail polish remover). Ketosis is seen most often in ewes in late pregnancy or in nannies in either late pregnancy or early lactation. Inadequate nutrition is the usual cause. However, ketosis has been associated with excessive consumption of mesquite beans.

Nitrate poisoning: Plants normally absorb nitrogen from the soil and incorporate it into plant proteins. However, nitrogen can accumulate in plants as excessive nitrate when the soil has high nitrogen levels from fertilizer or manure accumulation. Nitrates build up when rapid plant growth suddenly slows down, often because of sudden cloudy or cool weather or drought conditions.

Ruminant digestive systems convert nitrate into nitrite, which disrupts the red blood cells' ability to carry oxygen. At high levels, sudden death occurs, sometimes preceded by convulsions and kicking. Hence, the first sign of nitrate poisoning in livestock is often sudden death. Animals with nonlethal poisoning may breathe irregularly, grind their teeth, have dilated pupils or suffer delayed abortion 3 to 5 days after clinical signs in the dam.

Plants with more than 1.0 percent nitrate are dangerous; animals may die if they have consumed as little as 0.075 percent of their body weight in nitrate. Poisoned animals may be treated with methylene blue. The dead animals sometimes have dark brown, chocolate-colored blood and bluish or dark-colored mucus membranes, gums or tongues.

Plants that may cause nitrate poisoning include:

- Carelessweed, *Amaranthus* spp.
- Tansy mustard, *Descurainia pinnata*
- Kochia, *Kochia scoparia*
- Tumbleweed, *Salsola iberica*
- Nightshades, *Solanum* spp.
- Johnsongrass, *Sorghum halepense* and any other *Sorghum*

species or hybrids

- Goathead (puncturevine), *Tribulus terrestris*

Photosensitivity: A hypersensitivity to light producing a clinical sign termed "photosensitization." Animals may lick, rub or kick at themselves excessively. "Sunburned," reddened tissue usually appears on lightly haired or light-colored areas (muzzle, eyelids, teats, anus, vulva). A purplish band may appear near the top of the hoof, above the coronary band.

In severe cases, this is followed by blistering, and then swelling and excessive fluid accumulation (edema) which may progress to outward seepage. The edema that occurs in the face and muzzle of sheep gives the condition its common name "swellhead." Finally, in extreme cases, complete necrosis and sloughing of tissues of the skin, ears, lips, etc., occur. Secondary bacterial infections frequently develop.

There are two types of photosensitivity: primary and secondary or hepatogenous (liver-induced). Primary photosensitivity is caused directly by the plant toxin. Hepatogenous photosensitivity is more serious, because it results from a damaged liver. Toxins damage the liver, making it less able to eliminate the metabolic by-products of green plant digestion. These metabolites accumulate and cause photosensitization.

Plants that may cause photosensitivity include:

- Lechuguilla, *Agave lecheguilla:* hepatogenous
- Bishop's weed, *Ammi majus:* primary
- Hog plum, *Colubrina texensis:* hepatogenous
- Rain lily, *Cooperia pedunculata:* primary
- Wrights buckwheat, *Eriogonum wrightii:* primary
- Kochia, *Kochia scoparia:* hepatogenous
- Largeleaf lantana, *Lantana camara:* hepatogenous
- Sacahuista, *Nolina* spp.: hepatogenous
- Kleingrass, *Panicum colaratum:* hepatogenous
- Knotweed, *Polygonum* spp.: primary
- Dutchmen's breeches, *Thamnosma texana:* primary
- Goathead (puncturevine), *Tribulus terrestris:* hepatogenous

11

Prussic acid (cyanide or HCN) poisoning: A condition caused by an accumulation of plant metabolites (HCN complexes) in the leaves. In the animal, cyanide inhibits energy production in the cells, stopping their use of oxygen carried by the blood. Consequently, organ and tissue cells within the body are asphyxiated. Clinical signs usually occur within 15 to 20 minutes after the plant is ingested and progress very rapidly from excitability, salivation, rapid but difficult breathing, and muscle spasms to convulsions and rapid death. Blood is a cherry-red color at the time of death. If opened, the body cavity may smell like bitter almonds.

Treat cyanide-poisoned animals with sodium nitrite followed by sodium thiosulfate. It is critical to determine whether the problem is cyanide or nitrate before attempting to treat the animal, because part of the treatment for cyanide poisoning is nitrite, which can hasten nitrate poisoning. Most animals recover if they live more than 2 hours after clinical signs begin.

Plants that may cause prussic acid poisoning include:

- Mountain mahogany, *Cercocarpus montanus*
- Wild plum, *Prunus* spp.
- Johnsongrass, *Sorghum halepense* and any other *Sorghum* spp. or hyb.
- Queen's delight, *Stillingia* spp.

Polioencephalomalacia: A degeneration of the gray matter of the brain, "polio" is characterized by uncoordinated eye movements, blindness, abnormal posturing of the head and neck ("star gazing"), uncoordinated walking, convulsions and death. Rapid diagnosis is critical. Polio is treated with thiamine, and animals can be saved if treated early in the course of the disease. Conyza and excessive kochia consumption may cause polioencephalomalacia.

Vegetational Areas
of Texas

1.   Pineywoods
2.   Gulf Prairies and Marshes
3.   Post Oak Savannah
4.   Blackland Prairies
5.   Cross Timbers and Prairies
6.   South Texas Plains
7.   Edwards Plateau
8.   Rolling Plains
9.   High Plains
10.   Trans-Pecos

Plant Descriptions and Management Strategies

Guajillo
Acacia berlandieri

Guajillo is a nearly thornless shrub to small tree in the legume family. The plant height varies greatly, ranging up to 15 feet tall. The leaves are arranged like those of mesquite, but are smaller.

The sweet-scented, white to yellow flowers are clustered in dense groups. The plant produces a typical, flattened, bean-type legume fruit that is four to six times longer than it is wide.

Distribution and habitat

Guajillo grows on a variety of soil types but is most prolific on ridges and shallow soils. Found mainly in the Rio Grande Plains and southwest Texas, it is less common in the southern Edwards Plateau and Trans-Pecos regions. It also grows extensively in northern Mexico. Regions: 2, 6, 7, 10.

Toxic agent

Guajillo poisons sheep, goats and possibly cattle. The leaves contain excitatory amines, principally N-methyl-β-phenethylamine and tyramine. Overconsumption may cause a condition in sheep and goats known as guajillo wobbles, which may be followed by death.

The lethal dose of plant material for sheep and goats has not been determined. The toxic dosage is 15 times the animal's weight in leaves and fruit consumed over several months.

Livestock signs

Signs of guajillo poisoning include:

- Guajillo wobbles or limber leg in sheep and goats, an uncoordinated, rubbery action, usually in the rear legs
- Downed animals remain alert with normal appetite
- Death

At first, leg dysfunction appears only when animals are forced to move, but after several days the animal becomes unable to rise and may die. These signs are generally not apparent unless the animal has eaten guajillo almost exclusively for at least 9 months.

Research in sheep, goats and cattle has shown that the amine compounds can cause excessive release of stress hormones. As a result, certain reproductive functions (such as normal release of reproductive hormones, estrus, ovulation and testicular development) can be suppressed. Reduced pregnancy rates have been reported in nannies.

Integrated management strategies

Historically, guajillo has been considered a valuable browse plant that should not be grazed to the exclusion of other range forage. Recent evidence suggests that the plant has been overrated as a forage, because much of its nitrogen is of nonprotein origin and is poorly digested.

Supplemental feeding and reduced stocking rates lower the incidence of poisonings.

At the first signs of illness, move the animals to a pasture with more varied browse and especially herbaceous forage. Animals placed on an adequate ration may recover.

Managers should give animals alternatives to shrub foliage and avoid overstocking pastures.

Leaf ↗

Flower ↗

Whole plant ↓

Maple
Acer spp.

Maple trees can reach heights of up to 100 feet. The red maple *(Acer rubrum)*, often planted as an ornamental, is a beautiful tree with red flowers borne in the spring, before the leaves. Its three to five lobed leaves are 2 to 6 inches long and opposite each other on the stem, bright green above and whitish beneath.

Maples provide splendid fall color as the leaves turn shades of yellow through orange or red.

Distribution and habitat

Maple trees may be found throughout the entire eastern half of the United States and Canada, including all regions of Texas. Regions: 1, 2, 3, 4, 5, 6, 7, 8, 9, 10.

Toxic agent

The toxic agent of maple has not been identified. All species of maple should be considered potentially toxic to horses. Feeding studies have confirmed the toxicity of red maple. Most of the intoxications reported have dealt with red maple, but field cases are also reported involving consumption of silver maple with similar clinical signs.

Poisoning occurs when horses consume wilted or dry leaves equivalent to 1.5 to 3.0 grams of dry leaves per kilogram of body weight, or about 2 pounds of dry leaves for a 1,000-pound horse.

Fresh green leaves are not toxic.

Some animals have been poisoned after consuming bark; poisoning usually follows windstorms after which downed trees and limbs become available to the horses. Most cases of maple intoxication occur during the late spring, summer or early fall.

Livestock signs

Horses consuming maple may have hemolytic anemia and usually show clinical signs within 12 to 48 hours. The most prominent signs are:

- Yellow or brown mucous membranes
- Depression
- Anorexia
- Weakness
- Red or brown urine

Fatally affected horses usually die within 7 days. Sick animals may have a Heinz body anemia for 2 to 3 weeks before recovery.

Integrated management strategies

Never place limbs trimmed from maple trees in an area accessible to horses. Check horse pastures or paddocks containing maples for downed branches after a windstorm. Remove any damaged maple before it can be consumed.

↖ Flower

Whole plant ↗

Leaf ↓

19

Red Buckeye, Pale Buckeye

Aesculus pavia

Red buckeye is a shrub or small tree sometimes reaching 35 feet tall; its trunk reaches up to 20 inches in diameter and has smooth, gray or brownish bark. Large, deciduous, palm-shaped compound leaves are attached by petioles up to 6 inches long. The five serrate leaflets are up to 7 inches long and up to 2.5 inches wide.

The showy red flowers are displayed in erect clusters with tubular, five-lobed flowers and equal petals. A leathery capsule contains one to three large glossy brown seeds up to 1 inch in diameter. The western variety of this species is similar, but has pale yellow flowers.

Distribution and habitat

Buckeye is found in the eastern half of Texas and ranges east to North Carolina and Florida and as far north as Illinois. It is usually found in forests, along streams and on rocky hillsides in East Texas, and the shrub form is seen in improved bermudagrass pastures. The yellow-flowered variety usually grows along streams in canyons of the Edwards Plateau. Regions: 1, 2, 3, 4, 5, 7, 8.

Toxic agent

A glycoside called aesculin and/or a narcotic alkaloid is responsible for the toxicity of this plant. Buckeye has poisoned cattle, horses, sheep and swine as well as children.

Intoxication usually occurs in the spring when young tender leaves are present, especially in times of drought when other forage is short. Deaths in cattle have resulted from consumption of mature seeds off the ground.

Livestock signs

Many animals exhibit severe signs of intoxication within a few hours of consuming the plant. Clinical signs include:

* Uneasy, staggering gait
* Trembling
* Weakness
* Depression

Some animals suffer severe central nervous system depression, become comatose and die. Most cases, unless they are comatose, recover if further consumption is prevented. Because the onset of clinical signs is so acute, seed fragments are usually present in the rumen of cattle found dead after eating the seeds.

Integrated management strategies

Most fatal cases of buckeye poisoning occur when animals are forced to consume a large amount of plant material. Supplying adequate hay through the winter and into the spring during drought can prevent poisoning.

Animals that consume lethal amounts of the seeds usually have been introduced into previously vacant pastures in which seeds have accumulated. Keeping cattle in the pasture when the seeds are falling can prevent them from consuming a large amount of seed at one time.

← Flower

Leaf ↗

Whole plant ↓

Lechuguilla
Agave lechuguilla

Lechuguilla is a member of the same genus *(Agave)* as the century plant. Each plant consists of a crown bearing 20 to 30 thick, fleshy leaves 1 to 2 inches wide and 12 to 18 inches long. The leaves bend upward, have prickles on the margins and end in a sharp spine.

At 10 to 15 years old, the plant flowers once, then dies. The flower stalk produced at this time is 6 to 12 feet tall. New plants are formed from seed or by offsets from the parent plant.

Distribution and habitat

Lechuguilla is found in western Texas, southern New Mexico and south into Mexico. Prominent on dry hillsides and limestone hills, in dry valleys and in bordering canyons, it is especially abundant in the Trans-Pecos region of Texas. Regions: 7, 10.

Toxic agent

Sheep and goats are poisoned by lechuguilla most often under range conditions. Cattle are poisoned somewhat less frequently, although this plant can seriously threaten cattle during extended drought. Horses are not known to be poisoned.

Lechuguilla poisoning is thought to be the combined action of two photodynamic toxins, one of which is a hepatotoxic saponin. In experiments, sheep and goats fed as little as 1 percent of their body weight of lechuguilla leaf developed signs of photosensitization in less than a week, and died from the effects of liver and kidney damage in 1 to 2 weeks.

Livestock signs

Signs occurring under range conditions include:

- Listlessness and lack of effort to keep up with the flock or herd
- Decrease in water and food consumption and eventual anorexia
- Jaundice (yellow-tinted mucous membranes, eyeballs, skin, fat)
- Occasionally, port-wine-colored urine
- Photosensitization with swelling of the face and ears
- Progressive weakness and emaciation
- A short period of coma just before death

Postmortem examination may reveal swollen, greenish to black kidneys, a deep golden liver and yellow body fat.

Integrated management strategies

Remove animals exhibiting signs of poisoning from pastures containing lechuguilla. Place them in shade and give them good-quality feed and dry hay. Animals with severe jaundice usually die.

Because lechuguilla poisoning generally occurs when other, more desirable forage is lacking, any range management practice improving the range condition helps cut losses. Proper mineral supplementation, especially with phosphorus, is desirable. Remove plants along trails and in shipping traps, especially when hungry livestock are being trailed or held for shipping.

Herbicides do not control lechuguilla well. Grubbing provides good control, but is impractical for large areas.

Flower ↑

Whole plant →

↙ Leaf

23

Silktree, Mimosa
Albizia julibrissin

Silktree is a brittle-stemmed, short-lived tree that grows up to 40 feet tall. It is often broader than it is tall, with a flat top and an umbrella-like shape. The leaves are composed of several sets of small, pale-green leaflets on the short stems.

The showy flowers are borne in clustered heads at the ends of branches. The red or pink color comes from the multitude of stamens, which extend far beyond the other parts of the flower.

The flat, thin-walled, persistent seed pods measure about 0.5 to 1 inch across and 5 to 8 inches long, and contain numerous flat, brown seeds.

Distribution and habitat

Mimosa is a native of Asia and is widely planted as an ornamental across Texas. It has escaped and become naturalized primarily in the eastern third of the state. Regions: 1, 2, 3, 4, 5, 6, 7, 8, 9, 10.

Toxic agent

The legume (bean) contains a neurotoxic alkaloid that is responsible for the nervous signs and is thought to act as a pyridoxine (vitamin B_6) antagonist. Poisoning occurs when trees with green or mature pods are made available to cattle, sheep or dogs. They may gain access because of windfall or when trees have been pruned and the limbs with legumes are discarded where animals have access to them.

The lethal dose is about 1.5 percent of an animal's body weight in green or brown legumes containing seeds.

Livestock signs

Signs of poisoning occur 12 to 24 hours after intake of the legumes and include:
- Exaggerated response to stimuli
- Muscular twitching
- Labored respiration
- Salivation
- Convulsive seizures
- Death

Integrated management strategies

Pyridoxine (vitamin B6) injected intravenously can be an effective treatment even after seizures have begun.

Do not cut mimosa branches with seed pods and place them near livestock. Check pastures with naturalized trees after storms and remove branches with seed pods.

↖ Pod

Leaf and flower ↗

Whole plant ↓

25

Wild Onion
Allium spp.

There are 14 species and several varieties of wild onions in Texas. These biennial or perennial herbs have strong-scented (odor of garlic/onion), underground bulbs that give rise to long, narrow leaves.

The flowers are arranged in a terminal cluster attached to an unbranched stalk arising from the bulb between the leaves. Flower stalks may be 6 to 20 inches high with blooms of white, yellow, pink, red or purple. In some species, blooms are replaced by bulblets.

Distribution and habitat

Wild onions are widely distributed across the United States and are found during the spring in every region of Texas and in virtually every soil type. Regions: 1, 2, 3, 4, 5, 6, 7, 8, 9, 10.

Toxic agent

Onions contain N-propyl disulfide, which destroys red blood cells. Cultivated onions contain the same toxin and are often used as livestock feed. Many onions are usually required to cause poisoning; often pastures containing onions are heavily grazed without problems. The amount of toxin in the plants varies based on factors that are not well understood.

Cattle and horses are susceptible to onion poisoning, and cats are very sensitive to it. Sheep are more resistant, but have been poisoned by onions in some instances.

Livestock signs

Acute signs usually develop after a long-term intake of the plants and include:

• Jaundice (yellowish mucous membranes)

• Depression

• Anorexia

• Weakness

• Dark red or brown urine

Animals showing clinical signs have a strong onion/garlic odor on the breath as well as in the stomach/rumen contents. Sick or dead animals may have an onion odor without onion poisoning.

The clinical signs, especially the yellow color and dark urine, must be present to confirm the diagnosis. Diseases such as leptosporosis can also cause these signs.

Integrated management strategies

Most onion poisoning occurs when animals are fed waste onions. However, some cases occur when cattle or horses are forced to consume large amounts of wild onions. Good range management with adequate forage production will prevent wild onion poisoning.

26

↖ **Flower**

Bulblet ↗

Whole plant ↓

Whitebrush, Beebrush
Aloysia gratissima

Whitebrush is an aromatic shrub of the vervain family that grows from 3 to 10 feet tall. Leaves are narrow and pointed, pale beneath and 0.25 to 1 inch long. Leaves on flowering branches are small and smooth-edged, while leaves on other branches are larger and toothed. The tiny flowers vary from white to blue.

Distribution and habitat

Whitebrush is frequent to abundant in Central, West and South Texas and grows northwest into New Mexico and south into Mexico. In far West Texas, this plant is usually restricted to draws receiving extra runoff moisture and have deep soils. Regions: 1, 2, 3, 4, 5, 6, 7, 10.

Toxic agent

Horses, mules and burros are suspected to have been poisoned by this plant. The toxin, although unidentified, is known to be water-soluble.

Livestock signs

Signs of poisoning appear to be nervous in nature and include general weight loss. Feeding experiments conducted by the Texas Agricultural Experiment Station documented these signs in horses:

- Weakness
- Incoordination
- Prostration

These conditions developed in sequence beginning about a month and a half after access to whitebrush, ending in death about a week after appearance of definite nervous signs.

Integrated management strategies

A valuable honey plant, whitebrush has minimal value as browse for other classes of livestock or wildlife. Horses fed well and given a properly developed mineral and nutritional supplementation program are much less likely to consume whitebrush.

If desired, herbicides can control this plant. For broadcast applications, apply Spike 20P® at 1 to 1.5 pounds a.i./acre in late spring to early summer. Use higher rates on heavier clay soils. For individual plant treatments, apply Velpar L® at the rate of 4 ml per 1 inch stem diameter or 3 feet of plant height.

Double discing dense infestations and reseeding treated areas has reduced whitebrush by 90 percent.

↖ Flower

Leaf ↗

Whole plant ↓

Carelessweed, Pigweed

Amaranthus spp.

Carelessweeds are annual weedy herbs belonging to the amaranth family. Texas has 23 recorded species, which vary in growth forms from prostrate to branching upright. Carelessweed is often called pigweed because swine relish it. It bears inconspicuous flowers from June to November.

Distribution and habitat

Carelessweed abounds on disturbed sites—especially in barnyards with rich, moist soils throughout most of the United States. It is also a common weed in croplands. Regions: 2, 3, 4, 5, 6, 7, 8, 9, 10.

Toxic agent

Carelessweeds can accumulate nitrates from the soil to toxic levels. Environmental factors often influence nitrate accumulation. For example, nitrate poisoning is more likely to occur if the plant is growing in soils high in nitrogen, especially during drought.

Plants containing more than 1 percent nitrate are dangerous. The plant is also known to cause bloat. All ruminants are susceptible to nitrate poisoning.

Livestock signs

Animals with acute nitrate poisoning are often found dead with no previous history of illness. Less acute nitrate poisoning signs often occur in this order:

- Weakness
- Unsteady gait
- Collapse
- Shallow and rapid breathing
- Rapid pulse
- Dilated pupils
- Delayed abortion
- Coma
- Death
- Blood may appear chocolate brown at time of death.

Pregnant animals surviving acute nitrate poisoning may abort 3 to 5 days later.

Integrated management strategies

Many livestock relish carelessweed, particularly in its early growth stages. Although it is usually most dangerous during drought, poisonings have occurred at all growth stages and under a variety of conditions. The nitrate content of carelessweeds is significantly higher in the morning than in the afternoon.

Keep livestock off heavily infested pastures during early stages of plant growth and after sudden temperature changes. This plant remains dangerous in hay or silage.

Because livestock are most often poisoned when they are placed in a pen containing many carelessweed plants, focus herbicide or mechanical treatments on these areas. The plants' nitrate content may increase immediately after herbicide (especially 2,4-D) treatment; thus, keep livestock away until the plants have dried completely.

↖ Seedling

Flower ↗

Whole plant ↓

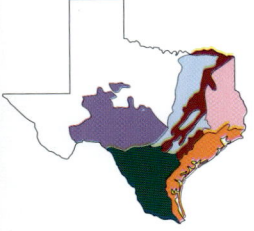

Bishop's-Weed, Greater Ammi

Ammi majus

Bishop's-weed is a showy, cool-season annual up to 3 feet tall. The oblong leaves may be up to 8 inches long and 5.5 inches wide.

The many white flowers are arrayed in an umbrella shape up to 3 inches across located at branch tips. Each flower gives rise to a small, oblong, rough fruit.

Distribution and habitat

This plant was introduced from the Mediterranean region and has been widely disseminated in the Western Hemisphere. It is found in east and south Texas, usually along roadsides, as it has apparently been included in some wildflower seed mixes.

In the past 15 years, bishop's-weed has become more widespread, and is now established in some roadside pastures. It is likely to spread further in coming years, posing a greater threat to livestock. Regions: 1, 2, 3, 4, 6, 7.

Toxic agent

Bishop's-weed contains a furocoumarin in all parts of the plant, but it is especially concentrated in the seed. The compound is photoactive, causing primary photosensitization in cattle, sheep and birds. All animals consuming the seed should be considered at risk.

Livestock signs

Signs in affected animals:

- Increased body temperature
- Photophobia (the animals shy away from light)
- Edema of the muzzle, ears, udder, scrotum and vulva
- Sunburn of light-colored skin
- Inflammation of skin

Thin-skinned areas and those having thin or no hair are often those most affected. Inflammation is followed by swelling, blisters, fluid seepage and sometimes sloughing of the skin.

In dark animals, the skin is not blistered or sloughed, but may become painful and thickened, with crusted hair.

Integrated management strategies

Primary photosensitization does not usually result in death. Some animals become severely scarred, and growth of young calves can be stunted because the cows will not let them nurse. Provide adequate water and feed in shade to prevent further sunburn.

Check pastures next to roads with bishop's-weed for flowering plants, which should be removed before seeds form.

↖ Leaf

Whole plant ↗

Flower ↓

33

Antelopehorn Milkweed

Asclepias asperula

Antelopehorn milkweed is an erect-stemmed plant growing to about 15 inches tall. Leaves are narrow, lance-shaped and about 3 inches long.

The flowers are greenish with distinctive purplish horns and are present from March to October. The fruit is a wrinkled pod containing silk-tufted seeds.

Distribution and habitat

Most abundant in western Texas, this plant has also been recorded in the northern, central and east central regions of the state. It ranges north into Nebraska and west into southern Utah and southeastern California.

It often abounds in open pastures, along arroyos, draws, bar ditches, trails and roadsides. Regions: 1, 2, 3, 4, 5, 6, 7, 8, 9, 10.

Toxic agent

The toxic agents involved are cardiac glycosides. Antelopehorn milkweed poisons all livestock, especially sheep. A toxic dose is generally considered to be 1.2 percent of the animal's body weight in green plant material.

Livestock signs

Signs of poisoning produced by most species of *Asclepias* differ only in degree. They include:

- First, profound depression, weakness and staggering
- Collapse, followed by frequent, intermittent muscular tremors
- Labored respiration, elevated temperature and pupil dilation
- Death after a comatose period of variable duration

Signs appear within a few hours of ingestion of a toxic dose, and death follows within a few days in most fatal cases.

Integrated management strategies

Animals dislike the taste of milkweeds and seldom graze them unless they are hungry and confined to milkweed-infested areas. Most losses are caused by overgrazing and drought.

The plant is most toxic before it matures, somewhat less as it dries. Antelopehorn milkweed retains enough toxicity to be dangerous in hay.

Although most animals die if they reach the convulsive stage of milkweed poisoning, some recover. Move them to shade, keep them quiet and give them plenty of food and water. No medicinal treatment is specified, but sedatives, laxatives and intravenous fluids may help.

↖ Flower

Pod ↗

Whole plant ↓

35

Broad-leafed Milkweed, Common Milkweed

Asclepias latifolia

More than 30 species of milkweeds have been recorded in Texas. Broad-leafed milkweed is noted for its robust nature and leaf size. A perennial, this plant has stout simple stems and four or more pairs of large thick leaves no more than twice as long as they are wide.

The flowers are greenish to white, giving rise to two to four smooth pods about 2 to 3 inches long from July to October.

Distribution and habitat

Broad-leafed milkweed is most common along trails and roadsides, less so in pastures. As with many weeds of low palatability, this species increases in heavily grazed pastures. It is frequent to abundant over much of the Trans-Pecos, the Plains and the central and western Edwards Plateau of Texas. It is found from Nebraska to Utah and west to Arizona. Regions: 7, 8, 9, 10.

Toxic agent

This plant poisons cattle and goats, but more often sheep. The toxic agents are cardiac glycosides. To be poisoned, cattle can eat as little as 1.0 percent of their body weight in broad-leafed milkweed; amounts as low as 0.15 percent have poisoned sheep and goats.

Broad-leafed milkweed is toxic in all growth stages, but is most toxic when immature. Cattle can generally graze frost-killed plants and not be poisoned.

Livestock signs

The signs produced by most species of *Asclepias* differ only in degree. They include:

- First, profound depression, weakness and staggering
- Collapse, followed by frequent, intermittent muscular tremors
- Labored respiration, elevated temperature and pupil dilation
- Death, after a comatose period of variable duration

Signs appear within a few hours of ingestion of a toxic dose, and death follows within a few days in most fatal cases.

Integrated management strategies

The best way to prevent losses from broad-leafed milkweed is to maintain good range condition. Removing plants along trails and in holding traps may prevent many losses, especially when hungry livestock are being trailed.

Avoid placing many animals where infestations are severe and forage is limited. Do not feed hay contaminated with milkweeds. Although no medicinal treatment is specified, sedatives, laxatives and intravenous fluids may help.

36

↖ Seedling

Flower ↗

Whole plant ↓

Horsetail Milkweed
Asclepias subverticillata

Horsetail milkweed is an erect-stemmed plant growing to 5 feet tall. The narrow leaves are in whorls of three or paired opposite, with margins rolled backward.

The greenish-white flowers give rise to pods 1 to 3 inches long from May to September. Seeds have tufts of long silky hairs.

Distribution and habitat

Abundant in western Texas, this plant has also been recorded in the South Texas Plains and Gulf Coast Prairie regions. It is frequent in northern Mexico and ranges into Arizona, Colorado and Utah. It often abounds in open pastures and along arroyos, draws, trails, roadsides and bar ditches. Regions: 6, 7, 8, 9, 10.

Toxic agent

The toxic agent involved is suspected to be the resinoid galitoxin. Horsetail milkweed has poisoned sheep, cattle, horses, chickens and turkeys. To ingest a toxic dose, an animal generally must eat 0.2 percent of its body weight in green plant material. A sheep may be killed by eating 2 to 3 ounces.

Livestock signs

The signs produced by whorled species of *Asclepias* are different from those produced by other milkweeds. Effects are on the nervous system rather than the cardiac system. They include:

- Staggering, incoordination, excitement
- Head tremors, muscle tremors, convulsions
- Star-gazing posture
- Depression, labored breathing, dilated pupils, progressing to death

Clinical signs appear within a few hours of ingestion of a toxic dose, and death follows from 1 to a few days in most fatal cases.

Integrated management strategies

Animals dislike the taste of milkweeds and seldom graze them unless they are confined to milkweed-infested areas. Most losses result from overgrazing and drought.

The plants are most toxic before maturing; somewhat less so as they dry. Horsetail milkweed retains enough toxicity to be dangerous in hay.

Although most animals die after reaching the convulsive stage of milkweed poisoning, some recover. Move them to shade, keep them quiet and give them plenty of food and water. No medicinal treatment is specified, but sedatives, laxatives and intravenous fluids may help.

↖ Pod

Flower ↗

Whole plant ↓

39

Peavine, Emory Loco
Astragalus emoryanus

Peavine is an annual legume with a slender taproot and stems growing close to the ground. The stems usually branch at the base and bear an odd number of sharp-tipped leaflets. The hairless, two-celled seed pods contain about a dozen seeds. Flowers are purplish, appearing from March to June.

Peavine is closely related to and can be found growing with other *Astragalus* species such as woolly loco and garboncillo. It is especially similar to the nontoxic *Astragalus nuttallianus* (Nuttall milkvetch).

Distribution and habitat

Peavine *(A. emoryanus)* is most abundant and causes the greatest threat in the limestone mountains of the Trans-Pecos region. It is also found in the red sandy soils of the Edwards Plateau and South Texas. Regions: 6, 7, 10.

Toxic agent

The toxic agents produced by peavine are miserotoxin and 3-nitro-1-propanol. It has recently been shown that some ruminal microorganisms detoxify these compounds, and the organisms' populations are increased if the animal receives small doses of toxin while on a high-protein diet.

Peavine is toxic to cattle, sheep and goats. Signs of peavine poi-soning were produced in sheep by feeding 1 to 2 percent of the body weight of peavine over 2 days.

Livestock signs

Under range conditions, signs resulting from chronic consumption of the plant are caused by loss of nerve function and may include:

- Momentary knuckling over of fetlocks
- Weakness in hindquarters
- Striking or rubbing rear hooves together
- Progressive hind-end incoordination
- Impaired vision
- Labored breathing, sometimes with a rasping voice

Integrated management strategies

Because peavine normally is short-lived, the usual management practice is to remove animals from pastures with severe infestations. Peavine is not a problem every year. During problem years, use of peavine-free pastures may be an alternative. Lightly stocking infested pastures may also limit poisoning.

At the first signs of poisoning, remove livestock from peavine pastures and place them in shaded pens with feed and water. Handle livestock suspected of consuming peavine quietly with as little stress as possible.

Herbicide applications are not usually economical. As with most annuals, mechanical soil disturbance may make the problem worse.

↖ Pod

Flower ↗

Whole plant ↓

Woolly Loco
Astragalus mollissimus

Woolly loco is a stout, many-branched perennial legume. Its leaves have 19 to 29 oval to oblong leaflets covered with fine, soft, short hairs. The thick, woody root gives rise to stems lying close to the ground.

The flower is purple, lavender or yellow, emerging in April through June.

Distribution and habitat

Woolly loco grows from south-western South Dakota to Texas and New Mexico. It is common in upland, mesa and mountain areas of the Trans-Pecos and Panhandle regions of Texas. Regions: 7, 8, 9, 10.

Toxic agent

Woolly loco is toxic to cattle, sheep, goats and particularly horses. Its toxin is swainsonine, which causes damage to the brain, liver, digestive organs, placenta and testes. The damage is reversible except in the brain. Locoism occurs in cattle and sheep after they have eaten about 90 percent of their body weight of the plant in a 2-month period. Ruminants usually must eat from 200 to 350 percent of their body weight over several months before death occurs.

In horses, about 30 percent of body weight consumption produces signs; about 75 percent may be fatal.

Livestock signs

Poisoning signs develop from the involvement of sensory and motor functions. In cattle, general signs include:

• Carrying the head a little lower than normal

• A vacant stare

• Trembling of the head

• Difficulty eating and drinking

• Abortion or weak/deformed offspring

Swainsonine is passed in the milk, possibly leading to unthriftiness in some suckling calves.

In horses, the time between the first clinical signs and death is much shorter than in cattle. The horse is listless, but on being stimulated, becomes excessively excited. Horses with chronic locoism rarely recover and are permanently dangerous to ride.

Integrated management strategies

Quickly move affected animals to locoweed-free pastures and place them on good feed. Management to reduce locoweed poisoning is most critical in early spring.

Livestock imported from areas where loco does not grow are the most susceptible to poisoning. Native animals generally avoid locoweed when good quality forage is available.

Maintaining good range condition and sound supplemental protein and mineral feeding programs are the best prevention against losses.

Locoweed dominating a particular area can be controlled chemically. Individual plant treatments or broadcast treatments with Grazon P + D® are effective. Locoweed becomes much more toxic and palatable after chemical application until it is completely dried up.

Pod ↗

Whole plant ↗ ↘

↖ Flower

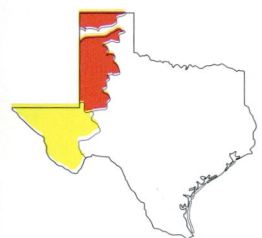

Garboncillo, Wooton Locoweed
Astragalus wootonii

Garboncillo is a much-branched annual legume with erect, hairy stems 3 to 12 inches long. The leaves are composed of nine to 19 leaflets, hairy beneath and smooth above.

The flowers are pink or purplish to white emerging in March to June. The most conspicuous part of the plant is the fruit, a large, one-celled, inflated pod.

Distribution and habitat

In Texas, garboncillo is generally restricted to the Trans-Pecos region. It is common in southern New Mexico, eastern Arizona and northern Mexico.

Garboncillo is most abundant in valleys receiving runoff water from the surrounding hills, as well as in bar ditches, along trails and around earthen tanks. Regions: 9, 10.

Toxic agent

The toxic principle is an alkaloid called swainsonine. Horses are particularly susceptible to garboncillo, although cattle, sheep and goats are also poisoned. Consumption of about 90 percent of the animal's body weight in garboncillo is required before cattle show signs of poisoning. As much as 200 to 350 percent of their body weight of the plant, eaten over a period of several months, may be required to kill cattle, sheep and goats.

Horses eating about 30 percent of their body weight in the plant show clinical signs, and consumption of about 75 percent of their body weight may be fatal.

Both dry and green garboncillo are toxic. Livestock readily eat the dead stems remaining after dieback at frost.

Livestock signs

Clinical signs develop from the involvement of sensory and motor functions. In cattle, general signs include:

- Carrying the head a little lower than normal
- A vacant stare
- Trembling of the head
- Difficulty eating and drinking
- Abortion

Swainsonine passes into the milk, possibly explaining the unthriftiness of some suckling calves.

The period from onset of signs to death is much shorter in horses than cattle. The horse is listless, but on being stimulated becomes excessively excited, even to the point of inflicting a fatal injury to itself. Horses with chronic locoism rarely recover and are dangerous to ride because of unpredictable and permanent behavioral changes.

Integrated management strategies

Because garboncillo is an annual, mechanically removing the plant around tanks, along roadways and in other hazardous sites is often effective. Proper stocking rates and grazing management practices can help reduce consumption. Sound range management practices can prevent garboncillo poisoning.

Livestock should receive proper supplementation of energy, protein, minerals and vitamins.

For small problem areas, treat individual plants with a 2 percent solution of Grazon P + D® applied directly to the leaves of the plant. For a large area, aerial or ground broadcast applications of 0.94 pounds a.i./acre of Grazon P + D® (3 pints) have given good results.

Flower ↗

Pod →

Whole plant ↓

Desert Baileya
Baileya multiradiata

An annual or weak perennial herb of the sunflower family, desert baileya grows to 1 to 1.5 feet tall. Leaves are arranged alternately along the stem and covered with woolly hairs.

The showy yellow flowers rise on long stalks from a leafy base. Desert baileya blooms from spring through late fall.

Distribution and habitat

Desert baileya is generally confined to desert regions from Texas to southern California and south into Mexico. Most often found on sandy/gravelly soils and dry plains and mesas up to 5,000 feet in elevation, the plant is also common on disturbed areas. Regions: 7, 10.

Toxic agent

The toxic agent is an unknown water-soluble compound. All parts of a green or dried plant are poisonous; flowers and seed heads are more toxic than leaves.

Sheep, goats and rabbits have been poisoned experimentally by desert baileya, although under range conditions only sheep are poisoned. Feeding trials suggest that 16 to 65 pounds of dry or green desert baileya are lethal to an adult sheep.

Although sheep eat baileya more readily when range feed is scarce, they have grazed it extensively when ample green grass was available. They appear to relish the flowers and seed heads.

Livestock signs

The first sign of poisoning in sheep is a frothy green salivation, followed by extreme weakness, a rapid heartbeat and trembling limbs. Under range conditions, poisoned animals may trail the flock with a stiff gait and show marked weakness. Other signs include:

- Rapid, pounding heart rate, audible without a stethoscope
- Trembling and loss of appetite
- Standing with back arched
- Lying down, unresponsive

Integrated management strategies

Graze problem areas with cattle only. Remove sheep and goats from infested pastures as soon as clinical signs are noticed, provide supplemental feed and good quality water and keep them calm.

Sheep poisoned by desert baileya may refuse to eat for a few days, but most regain their appetite and recover. Losses from desert baileya generally occur when other feed is short or when sheep are trailed through dense stands.

Because sheep like the flowers and seed heads, do not graze them in pastures with dense stands of desert baileya. Sheep losses from this plant in Presidio County were between $50,000 and $100,000 annually in 1958 and 1959.

Chemical and mechanical controls tend to be impractical. Losses can best be minimized through good livestock management.

Flower ↗

Whole plant ↓

Wild Indigo
Baptisia spp.

Wild indigo is a deep-rooted, perennial herb with bushy, branched, stout stems (often up to 3/8 inch at the base). Plants of some species may be as tall as 70 inches, but many specimens reach less than 15 inches.

The leaves are alternate with three deep lobes and two characteristic oblong structures attached at the base of the leaf stem.

The yellow or white flowers (rarely bluish) are in terminal or split spikes several inches long. In some species the spikes are erect and in others they hang like grape clusters. The fruit is a beaked pod containing two or more seeds. The dark gray or black leaf and stem color of the dead plants is a distinguishing feature of wild indigo.

Distribution and habitat

Various species of wild indigo are common in north central and eastern Texas. Texas species extend into Oklahoma, Louisiana, Arkansas and the eastern United States. They are usually found in sandy or sandy loam soil. Regions: 1, 2, 3, 4, 5, 8.

Toxic agent

Alkaloids present in wild indigo are believed to be responsible for its toxicity. These plants are very unpalatable and consumption is rare except when the animal is forced to eat it in hay, as has happened with horses. There are also very infrequent reports of cattle being affected by these plants.

Livestock signs

Signs of poisoning are those of a gastrointestinal upset and include:
- Colic
- Diarrhea
- Anorexia

In cases submitted to the Texas Veterinary Medical Diagnostic Laboratory, *Baptisia*-induced colic has been fatal to horses.

Integrated management strategies

Affected animals usually recover with supportive treatment after the contaminated hay has been removed. The black plant material is easily recognized in hay, which should not be fed in a manner forcing consumption.

Examine hay meadows for the plant before harvest. The individual plants can usually be removed mechanically.

Flower ↑ ↗

Whole plant ↓

49

Starthistles
Centaurea solstitialis, C. melitensis

Centaurea solstitialis, yellow starthistle, and *C. melitensis,* malta starthistle, are deep-rooted, branching annuals up to about 25 inches tall. The winged branches carry thick leaves and are narrow and smooth-edged near the tip and lobed at the base. Yellow flowers are borne at the ends of stems in the spring.

Malta starthistle is distinguished by its covering of dense white hairs and its thistle-like seedpod armed with sharp barbs.

Distribution and habitat

The starthistles are introduced invaders. Yellow starthistle covers millions of acres in California and is spreading in Texas. Malta starthistle has already become disseminated through most of the state.

Starthistles grow in disturbed, otherwise bare areas and as weeds in cultivated fields. Malta regions: 1, 2, 3, 4, 5, 6, 7, 8, 10; yellow regions: 4, 5, 7.

Toxic agent

The starthistle toxin is thought to be a sesquiterpene lactone. Toxicity occurs in horses after ingestion of 50 to 150 percent of the animal's weight in green plant material over a period of 1 to 3 months. Horses usually consume the young plant in early spring before stalk and spine growth. Yellow starthistle has been proven toxic to horses in experimental trials; malta starthistle is implicated only in case reports.

Livestock signs

Signs of poisoning in horses are attributed to brain damage and may include:

- Drowsiness
- Tongue flicking, lip twitching due to one-sided paralysis
- Chewing movements
- Difficulty eating and drinking
- Aimless walking

Integrated management strategies

Maintaining good range condition and sound supplemental protein and mineral feeding programs will help prevent poisoning. Even though malta starthistle is widespread across Texas and the southwest, there are few poisoning cases.

The greatest concern is probably the continuing spread of the more toxic yellow starthistle as a future problem in Texas. In areas of extreme infestations, chemical control strategies may be warranted.

Neither yellow nor malta starthistle is difficult to control with herbicides. Use general broadleaf weed compounds such as 2,4-D or

Grazon P + D®. Apply in the spring when there are 4 to 6 inches of growth and good growing conditions.

↖ Malta plant

Yellow plant ↗

Malta flower ↘

↙ Yellow flower

51

Mountain Pink
Centaurium beyrichii, C. calycosum

Two main species of *Centaurium* are found in Texas, both annual or biennial plants of the gentian family. The species grow to about 1 foot tall with many to few blossoms forming a rounded mass *(C. beyrichii)* or loose cluster *(C. calycosum)*.

The leaves are simple, opposite each other on the stem and attached directly to it. Pink flowers shaped like a five-pointed star are produced in late spring through summer.

Distribution and habitat

C. beyrichii is found in North Central to West Texas and into Arkansas; *C. calycosum* grows in Central and West Texas to Missouri west to Utah, Nevada and Arizona.

The main difference between these two species is their habitat. *C. beyrichii* usually grows in sun on dry, rocky limestone hills or in seeps on granite. It is most common in the Edwards Plateau. *C. calycosum* normally grows in moist habitats. Regions: 5, 6, 7, 8, 9, 10.

Toxic agent

The toxic principle of the *Centaurium* spp. is unknown. The plant is suspected to be poisonous to cattle, sheep and goats. Feeding of mountain pink produced illness in five goats; of those, four died. one of two sheep became ill.

The toxic dose is estimated to be between 0.5 and 1 percent of the animal's body weight consumed daily for several days. The plant was suspected of causing the death of bighorn sheep on the Black Gap Wildlife Management Area in Brewster County.

Livestock signs

Affected animals exhibit:

- Loss of appetite
- Signs of abdominal pain
- Diarrhea

Examination after death may reveal damaged tissue containing blood in the liver and kidney, and severe ulcers and inflammation in the rumen and abomasum.

Integrated management strategies

Mountain pink is relatively unpalatable to grazing livestock. Good grazing management practices that improve range condition can help reduce consumption of the plant. Offer adequate mineral and nutritional supplements.

Herbicides usually are unwarranted for minimizing toxicity problems from mountain pink.

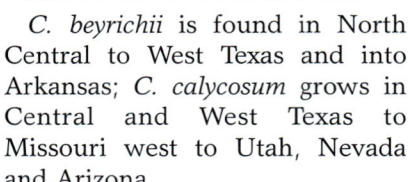

↖ *C. beyrichii* flower

C. calycosum plant ↗

C. beyrichii plant ↓

Buttonbush
Cephalanthus occidentalis

This plant is an upright, many-branched shrub or small tree, usually about 10 feet tall but ranging from 4 to 50 feet. Leaves are lance-shaped, up to 6 inches long and are attached two to four at each node.

The unique flowers are arranged spherically, forming white balls about 1 inch in diameter. They are soon replaced by dark-brown fruiting bodies that persist for several months. In the winter, this bush is reduced to branches tipped by the dark brown balls.

Distribution and habitat
Buttonbush is found in swamps, moist low-lying or irrigated areas and margins of streams throughout the state. Regions: 1, 2, 3, 4, 5, 6, 7, 8, 9, 10.

Toxic agent
The toxic agent has not been definitely established. However, early work from Europe has suggested a glycoside principle may be involved. This bush is very unpalatable and consumption and poisoning are unlikely. Many overgrazed pastures will have untouched buttonbush along streams. Cattle are thought to be the only species affected.

Livestock signs
Signs of poisoning are not well documented but may include:
• Vomiting
• Paralysis
• Muscle spasms

Integrated management strategies
The plant is not palatable and, therefore, good grazing management should prevent any problems. Severe starvation conditions must be present for cattle to consume buttonbush.

↖ Flower

Leaf ↗

Whole plant ↓

Mountain Mahogany
Cercocarpus montanus

Three varieties of mountain mahogany grow in Texas. Members of the rose family, they are identified most easily by the fuzzy, twisted awn attached to each fruit. The leaves are simple and often more than 1 inch long with heavily toothed margins.

This is an excellent browse plant for deer and elk.

Distribution and habitat

Mountain mahogany grows primarily on ledges and rims of rough terrain in western Texas.

Varieties can be found in the Trans-Pecos, Edwards Plateau and Plains regions in Texas. It also grows from Kansas to Arizona, South Dakota and Montana. Regions: 7, 8, 9, 10.

Toxic agent

Mountain mahogany contains concentrations of cyanogenic glycosides that under certain conditions are broken down to release cyanide. The toxins in the plant are usually, but not always, below the dangerous level.

Influences such as bruising, wilting, withering or drying of leaves appear to contribute to cyanide production. Wilted leaves from cutting the plant appear to be the most dangerous.

All domestic animals are subject to cyanide poisoning, although ruminants are the most susceptible.

Livestock signs

Cyanide is one of the most rapidly acting poisons. Signs of illness may start within 5 minutes from the time the animal begins eating the plant. Death may occur within 15 minutes or several hours.

Clinical signs generally occur in this order:

- Salivation and labored breathing
- Muscle tremors
- Incoordination
- Bright red venous blood
- Convulsions
- Death from respiratory failure

Integrated management strategies

Even plants containing a high level of cyanogenic glycoside may not poison livestock.

Because poisoning depends on the presence of free cyanide, anything preventing its development in the gastrointestinal tract can lessen or eliminate the danger of poisoning. Certain feeds, such as alfalfa hay and linseed cake, retard the production of cyanide and may prevent poisoning.

If signs are typical of cyanide poisoning, treat with sodium nitrite followed by sodium thiosulfate. Remove animals from infested pastures at the first signs of poisoning.

↖ Seed

Flower ↗

Whole plant ↘

↙ Leaf

Jimmyfern
Cheilanthes cochisensis

Jimmyfern is an erect evergreen with simple, fernlike leaves. The numerous leaflets are scaly beneath and have star-shaped hairs above. The leaves originate from a short, woody stem.

This is a "resurrection" plant: the leaflets roll up and become quite dry when moisture is lacking. They unroll and appear green and fresh after rain.

Distribution and habitat

Jimmyfern grows on rocky slopes and crevices, often closely associated with grasses characteristic of dry habitats.

Common in the Trans-Pecos, it also occurs in the Plains and Edwards Plateau regions of Texas. It extends into New Mexico, Arizona and south to Mexico. Regions: 7, 8, 9, 10.

Toxic agent

The toxic agent in jimmyfern is unknown. The poison is excreted in milk and is not destroyed when the plant dries.

Jimmyfern poisoning occurs in sheep, goats and cattle. A trembling reaction called "the jimmies" develops about 48 hours after animals are fed as much as 0.5 percent of their body weight in the fern and are exercised. Animals generally must walk 10 to 60 minutes to develop signs.

The danger of jimmyfern poisoning is greatest in wet years from mid-November through February when other forage is dry and the evergreen fern remains succulent and relatively palatable.

Livestock signs

Animals poisoned with jimmyfern show these clinical signs:
- Failure to keep up with the herd or flock
- Stilted, uncoordinated gait
- Arched back
- Violent trembling (the jimmies)
- Rapid heartbeat and breathing
- Prostration

Further exercise may bring on a fatal attack. Characteristically, in the last attack, the animal takes three or four stilted steps, drops, gasps a few breaths and dies almost immediately of respiratory paralysis.

Integrated management strategies

No specific treatment is known. Sheep and goats usually do not recover after eating a lethal dose of jimmyfern. Cattle tremble but usually live. Nonfatal cases require 5 to 19 days to recover. Under range conditions, it is important to leave poisoned sheep strictly alone during the danger period, because any excitement or exercise aggravates their condition.

58

Because most deaths occur in the winter, during this period supply enough forage or supplemental feed to animals in an infested pasture. Under severe conditions, move animals from pastures where jimmyfern is abundant in winter before additional losses occur.

Because of the habitat where jimmyfern typically grows, herbicidal control with spot applications is difficult and generally unjustified. Provide ample watering places in pastures where jimmyfern grows so that animals can drink conveniently.

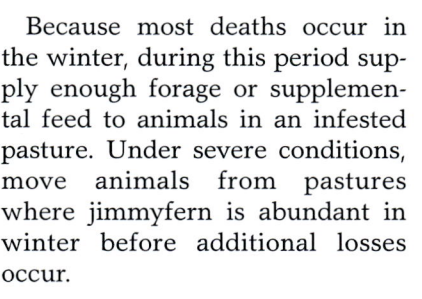

Leaf ↗

Whole plant ↓

Spotted Water Hemlock
Cicuta maculata

Spotted water hemlock is a stout perennial herb, 2 to 6 feet tall, arising from a tuberous base bearing fleshy or fleshy-tuberous roots.

Its short rootstocks have air cavities or chambers separated by cross-partitions containing a yellowish liquid that turns reddish brown when exposed to air. The stems are hollow except at the nodes and may be purple striped or mottled.

The two to three palm-shaped leaves are alternate, with stems wrapping around the main stem; they grow up to 15 inches long and 10 inches wide. The lance-shaped leaflets are 1 to 5 inches long and have saw-like margins.

Small white or greenish flowers are arranged at the ends of the stems in umbrella-shaped clusters.

Distribution and habitat

Spotted water hemlock is found in east, central and north Texas. Because it requires ample water, it is found only near streams, marshes, wet meadows and permanent springs. Regions: 1, 2, 3, 4, 5, 7, 8.

Toxic agent

The lower stalk and chambered rootstock of this plant contain most of its toxic alcohol, but hazardous concentrations can also occur in very young leaves. Mature and dried leaves are not toxic.

This fast-acting toxin can cause death between 1 and 8 hours after consumption in all animals, including humans.

Livestock signs

The toxic alcohol is a convulsant, and the clinical signs are the result of its action on the central nervous system. They include:

- Muscle tremors
- Salivation
- Grinding of the teeth
- Convulsions
- Death

Integrated management strategies

Water hemlock poisoning is not a significant livestock problem in Texas. Animals will die if they are forced to consume the young plant.

Do not pull plants for animals to consume, as this will make the highest concentration of toxin available (in the chambered rootstock) and can easily cause fatal poisoning.

↖ Leaf

Stem ↗

Whole plant ↘

↙ Root

19

Tobosagrass Ergot
Claviceps cinerea

Tobosagrass ergot is a toxic fungus that parasitizes the ovary of a developing tobosagrass flower. Infection occurs when the grass flower opens.

The fungus is a hard, pink or purplish structure that gradually replaces the grass grain. This fungal mass can be the same size as the grain it replaces or up to three or four times larger.

Distribution and habitat

Tobosagrass grows on dry, rocky slopes, dry upland plains and plateaus, and heavy clay soils.

It is found from central Texas west to southern Arizona and south into Mexico. The ergot fungus can infect plants wherever they grow. Regions: 6, 7, 8, 9, 10.

Toxic agent

Ergot toxicity is caused by a variety of alkaloids and, sometimes, tremorgens.

Few experiments have been conducted to determine the exact amount of ergot needed to produce poisoning in cattle, or the amount of time required for clinical signs to appear.

Livestock signs

Tobosagrass ergot poisoning in animals other than cattle is relatively rare. Two types of ergot poisoning

have been described: nervous ergotism and gangrenous ergotism. The fungus affecting tobosagrass typically produces only the nervous type.

Clinical signs of nervous ergotism include:

- Hyperexcitability
- Uncontrollable muscle tremors
- Incoordination
- Falling when forced to exercise
- Inability to regain feet

Animals usually recover when they are removed from ergotized pastures, unless they go down in the sun and die of exposure or lack of water.

With tobosagrass, the signs are only rarely accompanied by gangrenous injury. Gangrenous ergotism usually affects parts of the body having the poorest blood supply, such as the feet, legs, tail and ears, any or all of which may drop off. Cattle so affected can walk without hooves, apparently without pain.

Abortion is also associated with gangrenous ergotism.

Integrated management strategies

Cattle do not dislike ergot, and while grazing a pasture they may at first eat only the grass heads, and therefore consume a concentrated amount of ergot. If signs of poisoning occur, move cattle from the pasture immediately.

If handled properly, animals with signs of nervous ergotism usually recover within 5 days to 2 weeks.

If possible, avoid pastures where ergot has been identified.

Ergot body ↑

Whole plant ↓

63

Dallisgrass Ergot
Claviceps paspali

Dallisgrass *(Paspalum dilatatum)* seed heads are often infected with the fungus *Claviceps paspali*. Fungal spores germinate in the flower, grow in the premature seed and produce honeydew that is transferred to other seed heads by insects. In an infected flower, a fungal body or sclerotium forms instead of a seed. The sclerotium is round and up to ⅛ inch across with a cream-colored center. Its outer coat may vary from white to orange, red or black because of other fungi growing on the ergot body.

Many seed heads include normal seed, honeydew and sclerotia. This fungus also infects and produces toxins in other *Paspalum* species such as brownseed paspalum and bahiagrass.

Distribution and habitat

Dallisgrass may be found in all areas of Texas, but widespread ergot infection is usually limited to the eastern half of the state. The high humidity and soil moisture required for ergot production also occur in limited areas along streams and in canyon floors in west central and West Texas. Regions: 1, 2, 3, 4, 5, 6, 7, 8, 9, 10.

Toxic agent

Sclerotia on *Paspalum* grasses contain paspalitrems, termogenic

mycotoxins responsible for dallisgrass staggers. Most cattle poisonings result when the cattle eat mature seed heads in the pasture; calves are known to selectively eat seed heads with honeydew.

Horses are not usually poisoned unless they consume *Paspalum* hay that contains ergot.

Livestock signs

Poisoned cattle and horses demonstrate similar signs, including:

- Hyperexcitability
- Uncontrollable muscular tremors
- Incoordination
- Falling when forced to exercise
- Inability to regain feet

Cattle usually recover when they are removed from ergotized pastures unless there is misadventure, such as falling headlong into water or limb breakage, or if they go down in the sun and die of exposure or lack of water.

Horses with nervous ergotism tend to be destructive, often injuring themselves, sometimes requiring euthanasia.

Integrated management strategies

To treat for nervous ergotism, remove the source from the animals' diet. Some severely poisoned horses have recovered after

64

a few days in a padded surgical recovery room. Cattle almost always recover if they are moved to shade, fed and watered.

To prevent poisoning, managers must be able to recognize ergot-infected seed heads and prevent livestock from consuming them. Remove the seed heads by mowing before cutting for hay or grazing the pasture.

If grazing is continuous, most seed heads are consumed before the toxin is produced. Unrolling potentially hazardous round bales can leave many of the ergot bodies on the ground, where animals are less likely to eat them.

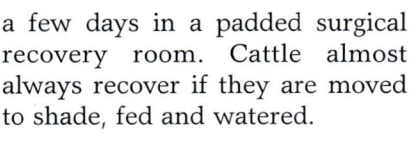

Ergot body ↗ ↘

↙ Whole plant

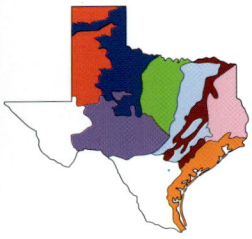

Ryegrass Ergot
Claviceps purpurea

The ergot of ryegrass *(Lolium* spp.) is a fungal body or sclerotium often found in many cereal grasses. It is a purplish black, banana-shaped, 0.25- to 0.75-inch-long body replacing a grain in the seed head.

The ergot body drops to the ground with the seed, lies dormant through the winter and produces millions of spores in the spring.

Distribution and habitat

Annual ryegrass is found throughout the eastern half of Texas and in the Panhandle. It grows as a cool-season forage and as a volunteer in disturbed areas. Severe contamination by ergot is usually restricted to the eastern third of Texas, where rainfall is more abundant. Regions: 1, 2, 3, 4, 5, 7, 8, 9.

Toxic agent

The ergot alkaloids in the sclerotia cause smooth muscle contraction, which limits blood supply to the extremities. In the United States, cattle are most often affected.

If consumption occurs in winter, dry gangrene may result; in the summer, animals exhibit heat intolerance and poor performance.

The ergot of mature annual ryegrass fractures easily from the seed head; most poisoning occurs when animals consume harvested seed or grain screenings. Hay seldom causes problems because it is usually harvested before the seed heads, sclerotia and toxins fully develop.

Livestock signs

Signs of gangrenous ergotism vary depending on dose and the time of year and may include:
- Loss of the tips of the ears
- Loss of the tip of the tail
- Loss of one or more feet
- Standing in shade or water
- Poor weight gain
- Abortion

Integrated management strategies

Prevent poisoning by harvesting ryegrass during or before the early dough stage of seed development. Inspect ryegrass seed and seed screenings for ergot sclerotia before feeding, and do not place animals in mature, heavily infected monoculture pastures.

Ergot body ↑

Whole plant ↓

Hog-plum,
Texas Colubrina
Colubrina texensis

Hog-plum is a many-branched, 4- to 6-foot-tall shrub. Some branches form slender, weak spines. The leaves are small, alternate, borne on stems and somewhat oval.

Small, yellow flowers are solitary or on short, twig-like shoots. The small, brownish fruit is a hard, three-celled, spherical capsule.

Distribution and habitat

Hog-plum is found in gravelly arroyos and on rocky hillsides in southwestern Texas. Regions: 2, 5, 6, 7, 8, 10.

Toxic agent

The toxicity of this plant is questionable, and any potentially toxic agents are unidentified. The fruits and seeds are the plant parts of concern.

Livestock signs

Hog-plum is browsed by all species of livestock. Controlled experimentation has not been performed to establish its potential toxicity.

Field reports indicate that sheep may have been poisoned after ingestion of fruits and seeds. The condition is similar to lechuguilla poisoning, with clinical signs including:

- Photosensitization, with swelling of the head and ears
- Jaundice (yellow discoloration of mucous membranes, eyeballs, skin, fat)
- Death

Integrated management strategies

The plant is usually not a problem for livestock producers. Avoid placing sheep in pastures with heavy populations of hog-plum when the plant is in fruit or seed stages and when other forage is not readily available. Observe livestock closely and remove animals from infested pastures at the first signs of poisoning.

↖ Leaf

Fruit ↗

Whole plant ↓

Poison Hemlock
Conium maculatum

Poison hemlock is a biennial of the parsley family. It has stout, erect, hollow stems that may be purple streaked or splotched and may grow to up to 10 feet tall.

Leaves can be 6 inches wide and 12 inches long, with many oval to broadly oval leaflets opposite each other. The leaf stems clasp the main stem at their junction.

White flowers are arranged in umbrella-shaped clusters.

Distribution and habitat

These plants are found in dense stands in roadside ditches and stream banks in the southern half of Texas. Regions: 2, 3, 4, 6, 7, 10.

Toxic agent

Poison hemlock contains pyridine alkaloids. The stems and leaves are the most toxic parts of the plant. Cattle and swine are the species affected most often. This plant is hazardous to humans and was used in political executions in ancient Greece (Socrates).

Cattle seldom graze the plant, but may be poisoned by it in hay or green chop. The roots or young leaves may poison swine. Hay containing poison hemlock is considered hazardous.

Livestock signs

Signs of acute poisoning occur within a few hours of consumption; these include initial stimulation followed by progressive central nervous system depression.

Stimulation
- Nervousness
- Muscle tremors
- Incoordination
- Salivation
- Colic

Depression
- Partial paralysis
- Slow heart rate
- Low body temperature
- Slow respiration rate
- Coma
- Death

Low-level intake during a specific period of gestation (50 to 75 days in cattle) may cause birth defects. "Crooked calf disease" may include cleft palate and skeletal deformities. The amount of plant needed to produce birth defects may not be enough to cause illness in the dam.

Integrated management strategies

Do not cut hay or green chop from areas containing this plant. This is especially of concern when roadsides are used to harvest low-quality hay during drought.

Severely poisoned animals may be given stimulants and supportive care.

↖ Stem

Flower ↗

Whole plant ↘

↙ Leaf

71

Coulter Conyza,
Horsetail Conyza
Conyza coulteri, C. canadensis

Conyza coulteri and *C. canadensis* are herbaceous, 3- to 6-foot-tall annuals of the sunflower family. The leaves of *C. canadensis* are long and narrow, while those of *C. coulteri* are lobed and rounded on the ends.

The many-flowered heads grow in leafy spreading branches or clusters, becoming feathery white after flowering.

Distribution and habitat

Coulter conyza is confined to the western third of the state, while horsetail may be found in all areas of Texas. Both species are commonly found in disturbed sites. Regions: 1, 2, 3, 4, 5, 6, 7, 8, 9, 10.

Toxic agent

The toxic agent of these plants is unknown. Coulter conyza has poisoned sheep, goats and cattle in experimental trials. During drought, this plant has been responsible for serious losses of cattle in the Trans-Pecos.

Horsetail conyza has not been fed experimentally, but cattle and goats have been lost after consuming the plant. Both result in polioencephalomalacia.

Livestock signs

Clinical signs associated with poisoning by these plants are related to brain damage and may include:

- Walking in circles
- Hyperexcitability
- Muscle tremors
- Apparent blindness
- Coma
- Death

If begun early enough, treatment with thiamine may reverse the condition.

Integrated management strategies

Both species of conyza are unpalatable, and livestock eat them only when forced. Coulter conyza is normally a problem in drought when the plants are young and forage is limited. Horsetail is usually consumed only after it is treated with the herbicide 2,4-D; do not allow animals access to treated plants until the plants are dead and dry.

These plants readily invade disturbed areas and often increase after mechanical brush control has disturbed the soil. Coulter conyza has increased where Spike 20P® was used to control woody plants.

72

↖ **Coulter flower**

Horsetail plant ↗

Coulter plant ↘

↙ **Coulter seedling**

Rainlily
Cooperia pedunculata

Rainlily grows from a black underground bulb up to 1.5 inches tall, conical when young and becoming flattened to round.

The 0.5-inch-wide linear leaves become narrower toward the end and may be up to 14 inches long depending on rainfall and soil moisture. Each leaf ends with dry, desiccated material measuring from less than 0.25 inch to as much as several inches long.

A flower stalk up to 8 inches high with a single white bloom up to 2 inches across appears a few days after a rain. The plant may or may not have leaves when the flower appears.

Distribution and habitat

Rainlily is found in east, central and southwestern Texas. Depending upon the region, it may be found along streams, in valleys or on hillsides. Regions: 1, 2, 4, 6 and 7.

Toxic agent

The toxic agent of rainlily has not been identified, but it is found only in the dead leaf tips of plants growing in a limited geographical area. Biological examinations of specimens throughout the range indicate that plants in and near DeWitt, Gonzales and Caldwell counties are highly photodynamic while those from other areas have little or no photoactivity.

Photosensitivity occurs primarily in the fall, but also during late spring or summer, usually within 10 days after a rain. This suggests that some microbiological activity on the dead leaf material is responsible for the activity. Severe outbreaks of photosensitization occur when rain falls on a large amount of dead leaves.

Livestock signs

Rainlily causes primary photosensitization, and poisoning results in loss of production but usually not death. The clinical signs are those of sunburn and include:

- Photophobia (animals try to stay in shade)
- Sunburn of light-colored skin
- Crusting and cracking of light colored skin
- Sloughing of skin

This plant most often affects cattle in areas where the plants have high photoactivity. Severe outbreaks have included white-tailed deer. Deer and black cattle may become blind, and their eyes may turn cloudy and blue.

Integrated management practices

Place affected animals in the shade and give them adequate feed and water. Give supplemental feed to calves nursing severely

affected cows, as they are often not allowed access to painful udders.

Most ranchers with large rainlily populations in DeWitt County have only black cattle, which prevents the majority of problems in most years.

Flower ↗

Bulb ↘

← Whole plant

Golden Corydalis
Corydalis aurea

Golden corydalis is a spreading, yellow-flowered member of the poppy family. It is sometimes referred to as scrambled eggs. Its stems are pale or whitish and leafy. The leaves have numerous small segments.

The yellow flowers, irregularly shaped and spurred at the base, occur from February to May. They appear in loose clusters at the ends of the branches. The seeds are black and shiny.

Distribution and habitat

The plant is distributed widely in the western half and central parts of Texas. It is also found north to South Dakota, west to Utah and in southern Nevada and Arizona.

Golden corydalis often grows in disturbed areas, along stream banks, in open woods and in sandy soils throughout Texas. Regions: 5, 7, 8, 9, 10.

Toxic agent

This plant reportedly contains up to 10 alkaloids. Sheep like this plant, and eating as little as 2 percent of their weight can cause clinical signs. Less than 5 percent can be fatal.

Although this plant reportedly poisons cattle and horses, they generally are much more resistant than sheep. Goats are least susceptible, cattle intermediate, and sheep are most susceptible.

Livestock losses from this plant are much lower in Texas than in other states, such as Arizona.

Livestock signs

Cattle and sheep clinical signs are similar and often appear within minutes and usually within a few hours of consuming corydalis. They include:

- Uneasiness
- Twitching facial muscles
- Rapid respiration
- Staggering and falling into convulsions

Downed animals make running motions with their feet. Diarrhea and bleating or bawling are common. Throughout the period of clinical signs, the animals bite at nearby objects.

Affected animals usually experience several cycles of these fits, between which they exhibit normal behavior. In lethal cases, breathing and heart action slow, and after successive convulsive periods, the animal dies. Animals not lethally poisoned usually recover quickly and uneventfully.

Integrated management strategies

Golden corydalis has not been a serious problem in Texas. Range management practices promoting improved range condition help reduce losses to this plant. Proper supplemental feeding programs also help.

For extreme populations, graze infested pastures with livestock least susceptible to corydalis poisoning.

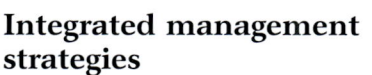

↖ Leaf

Flower ↗

Whole plant ↓

Bermudagrass
Cynodon dactylon

Bermudagrass is a bluish-green, perennial sod grass having both rhizomes and long stems or runners that take root at the nodes. The internodes are flattened.

The seed head has three to six purple spikes and resembles a bird's foot. Along one edge of the seed are hairs visible only with a microscope.

Many varieties and hybrids of bermudagrass have been planted throughout the state.

Distribution and habitat

This plant or its hybrids are found in all regions of Texas. It is the most important warm-season perennial grass planted for improved pastures in the state. Regions: 1, 2, 3, 4, 5, 6, 7, 8, 9, 10.

Toxic agent

The toxic agents in bermudagrass are not known. Two of the toxic syndromes associated with this plant are related to fungal growth.

The first syndrome, bermudagrass staggers, results from an unknown tremorgenic mycotoxin produced by an endophyte (a fungus within the plant). Seen only in cattle, the staggers syndrome may result from the consumption of the stable toxin in hay or from pastures, usually in the fall.

The second syndrome produces liver damage and secondary photosensitization when the grass being grazed has a heavily moldy thatch beneath it.

A third disease, fog fever or pulmonary adenomatosis, occurs when extremely lush bermudagrass is grazed. This grass contains unusually high levels of the amino acid tryptophan, which is converted by rumen microbes to the lung toxin 3-methyl indole.

Livestock signs

Bermudagrass staggers is similar to dallisgrass staggers except that there is less hyperexcitability. Signs become pronounced upon exercise and there may be:

- Head bob
- Muscle tremors
- Incoordination
- Collapse when forced to make rapid movement
- Inability to regain feet
- The condition worsens if the animal is assisted.

The staggers syndrome is reversible, and most cattle recover unless there is a fatal accident.

Animals with liver damage caused by bermudagrass have photosensitization. Those with fog fever have severe breathing difficulty, and can die from lack of oxygen.

Integrated management strategies

Remove cattle with staggers from the toxic hay or pasture, supply them with feed and water and allow them to remain as quiet as possible. Complete recovery may take up to 3 weeks.

Hay and grass in the pasture should be destroyed.

Place animals with secondary photosensitization in shade, feed them sun-bleached hay having no green color and treat them symptomatically. Feeding monensin, a feed additive that shifts the population of rumen microbes, may prevent fog fever problems.

Vacate pastures causing the problem for 14 days.

Seed head ↗

Whole plant ↓

Jimsonweeds, Thornapples
Datura spp.

Several species of jimsonweeds grow in Texas, all appearing very similar. They are coarse-looking, ill-scented, herbaceous annual weeds of the nightshade family.

The very distinctive flowers are large, showy, erect, solitary and white or purplish and may appear from April to October. Flowers grow in the leaf axils. Leaves are alternate, simple, hairless and toothed.

Except that they are larger, the seedpods resemble those of cocklebur, being a spiny capsule up to 2 inches long.

Distribution and habitat

Jimsonweeds are distributed widely throughout Texas and the United States. They normally grow in rich soils on disturbed sites, waste places, old fields and in open areas. Regions: 1, 2, 3, 4, 5, 6, 7, 8, 9, 10.

Toxic agent

Toxicity results from tropane alkaloids (atropine, scopolamine, hyoscyamine). Livestock and people can be poisoned by eating any part of the plant, including the seeds.

Consumption of as little as 10 to 14 ounces of the plant, or less than 0.1 percent of the body weight of the animal, can kill cattle.

Many humans have been poisoned by eating the seeds and unripened seed pods and flowers.

The plants have been ingested for their hallucinogenic effects.

Livestock signs

Signs of poisoning are similar for both humans and livestock. Clinical signs include:

- Intense thirst
- Distorted vision
- Uncoordinated movement
- High body temperature
- Rapid and weakened heartbeat
- Dilated pupils
- Convulsions
- Coma and death

Integrated management strategies

No medicinal treatments are specified for poisoned livestock, although stimulants such as pilocarpine and physostigmine have been used. Some poisoned animals may recover if they are hand-fed a suitable diet and provided good quality water.

Jimsonweeds are extremely unpalatable to livestock and poisoning occurs only rarely. When it does occur, it is usually because hungry livestock are confined where plants are found.

Grubbing or herbicides easily control this plant. Exercise caution when treating jimsonweed with 2,4-D, as it makes the plant more palatable.

↖ **Whole plant**

Pods ↗

Flower ↓

81

Larkspur
Delphinium spp.

The larkspurs are perennial herbs growing 1 to 3 feet tall. The leaves are deeply divided from a single point into numerous fine, narrow segments and are usually confined to the bottom half of the plant.

Showy, light blue, blue, purple or white flowers are arranged along the top of the erect stalk. The distinguishing characteristic of the flower is the prominent backward trailing "spur."

Distribution and habitat

One or more species of *Delphinium* may be found in open pastures, hillsides or valleys of all vegetational areas of Texas. Regions: 1, 2, 3, 4, 5, 6, 7, 8, 9, 10.

Toxic agent

More than 40 different diterpenoid alkaloids have been identified from *Delphinium* spp., but most of these are from the tall larkspurs of the mountains in the western United States. Toxicity varies greatly among species, as does the concentration of toxins.

Usually, alkaloid concentrations are highest in the spring after flowering and gradually decrease as the plant matures. All parts of the plant should be considered toxic. All species of livestock may be affected, though cattle are the most susceptible.

Texas larkspur species are small plants with scant foliage, and poisoning is uncommon. Despite this, all larkspurs should be considered to be potentially toxic.

Livestock signs

Larkspur primarily affects the neuromuscular system and consequent signs of poisoning include:
- Salivation
- Arched back
- Stiff gait
- Collapse, followed by struggle to regain feet
- Muscular twitches
- Sudden death

Within 3 to 4 hours of consumption, death can occur from either paralysis of the respiratory system or asphyxiation caused by bloating or vomiting.

Integrated management strategies

Larkspur poses the greatest risk to livestock in the spring. Losses in heavily infested areas may be reduced by keeping cattle out of pastures until after the plants have flowered and gone to seed, or by grazing sheep, which are more resistant, before cattle.

Because treating larkspur with some herbicides increases its palatability, do not use treated pastures for grazing until the affected plants are dead.

82

Leaf ↗

Whole plant →

Flower ↘

← Whole plant

83

Tansy Mustard
Descurainia pinnata

Tansy mustard is an annual cool-season forb growing to 2 feet tall. It is usually single-stemmed, leafy and covered with fine, gray hairs.

Leaves are placed alternately along wavy stems, with each divided into many small segments.

Flowers vary from yellow to whitish, occurring in long clusters at stem ends. Plants flower in February through May.

The very distinctive fruits are long, round, slender, two-celled capsules filled with many small, waxy seeds.

Distribution and habitat

Tansy mustard is distributed widely throughout the southern and western United States up to 7,000 feet in elevation. Heavy stands may form on dry, sandy soils in arid areas. Abundance increases after moderate or heavy winter rains in the arid southwest. Regions: 2, 3, 4, 5, 6, 7, 8, 9, 10.

Toxic agent

The toxic agent is unknown. Large quantities of the plant must be consumed before poisoning occurs. Tansy mustard also accumulates toxic levels of nitrate.

Livestock signs

Cattle are the only kind of livestock reported to be poisoned. The first clinical sign is partial or complete blindness (blind staggers). Animals wander aimlessly until exhausted, or may stand pushing their head against a solid object for hours.

Next, or along with blindness, comes an inability to use the tongue or to swallow (paralyzed tongue). Cattle may stand at water unable to drink, or try unsuccessfully to graze.

Integrated management strategies

A simple and effective treatment is to administer 2 to 3 gallons of water (with nourishment such as cottonseed meal if the animals are seriously weak) twice daily by stomach tube. With this treatment, clinical signs gradually disappear.

Tansy mustard is relatively non-toxic, so moderate amounts may be desirable. Furthermore, stands thick enough to lead to the heavy consumption necessary for poisoning do not appear every year. Because of this, herbicidal control is not recommended except where dense stands occur near watering and holding facilities and other areas of high livestock use.

Good range management practices and grazing a mixture of cattle, sheep and/or goats may help prevent excessive intake by cattle.

↖ **Whole plant**

Leaf ↗

Flower ↓

85

Texas Persimmon
Diospyros texana

Texas persimmon, also called Mexican or black persimmon, is normally a shrub or small tree less than 15 feet tall. However, some specimens along the upper Texas coast may reach 50 feet tall. The compact wood is almost black, and the gray, slick bark is thin.

The oval leaves, rounded at the tips, have small, fine hairs on the lower surface. The fruit contains three to eight seeds, can measure up to 1 inch in diameter and is green, turning to black when ripe.

Distribution and habitat

Texas persimmon is primarily found in the western two thirds of the state in rocky open woodlands, arroyos and on open slopes. In some pastures in central Texas, it may be one of the predominant invading woody species. Regions: 2, 3, 4, 5, 6, 7, 8, 9, 10.

Toxic agent

The toxic agent in persimmon is unknown, and the information in this section is based on observation rather than experimental studies.

During periods of drought, there have been years when the persimmon crop was heavy and grass was sparse. Cattle consuming large amounts of ripe or ripening fruit have had problems.

Livestock signs

Persimmon poisoning does not cause death, but it does result in poor performance. The clinical signs are:

- Black diarrhea
- Colic
- Weight loss

At times, the level of intake is so great that persimmon seeds are about the only solid material in the feces.

Integrated management strategies

Cattle regain their weight when given adequate nutrition after the fruit is gone. As this usually occurs in drought, expensive supplemental feeding can be required.

Prevention is best and can be accomplished by moving cattle to pastures with fewer plants when the fruit begins to ripen.

Individual plants of Texas persimmon may be controlled with a basal stem treatment mixture containing 25 percent Remedy® and 75 percent diesel fuel oil to thin out heavily infested pastures. Apply the mixture to the bottom 12 inches of the stem down to the soil surface in the spring after the leaves mature but before June 15.

← Fruit

Leaf and flower ↗

Whole plant ↓

87

Inkweed,
Thickleaf Drymary
Drymaria pachyphylla

Inkweed is a smooth, hairless, short-lived annual with blunt-pointed, circular leaves usually about as wide as they are long.

The plant grows close to the ground in a circular pattern 5 to 10 inches in diameter. Small flowers are produced in the leaf axils.

Distribution and habitat

Inkweed grows on sites with sparse vegetation, most often on heavy, alkaline clay soils and in low areas subject to occasional flooding. It is generally not locally abundant except on disturbed sites.

Inkweed is found in western Texas, southern New Mexico, west to southeastern Arizona and south into Mexico. Region: 10.

Toxic agent

The plant poisons cattle, sheep and goats. Its toxic agent is unknown. All parts of dry and green plants are toxic.

Most poisoning occurs on overgrazed ranges. Plants are most often grazed in the early part of the day when they are swollen with water and more upright.

Feeding experiments have shown that a lethal dose is 0.6 percent of body weight for a sheep; 0.4 percent for a cow and 0.9 percent for a goat.

Livestock signs

Signs appear 18 to 24 hours after a toxic dose is ingested. Death usually occurs less than 2 hours after the first clinical signs occur. Signs appear in this order:
• Loss of appetite
• Diarrhea
• Arched back and tucked up abdomen
• Coma
• Death

Integrated management strategies

Inkweed is very unpalatable to all classes of livestock. Poisoning generally occurs only when other forage is limited.

Range management practices that improve range conditions and increase forage diversity help reduce losses to inkweed. Take special caution in drought years.

↑ Leaf

Whole plant ↓

69

Wright Buckwheat
Eriogonum wrightii

Wright buckwheat is a low, highly branched, perennial shrub that grows from a robust taproot and a reddish woody base. The oblong leaves are about 1 inch long and are scattered on many stems.

When in flower (August to October), the plant appears as a white bouquet that turns reddish-orange in cool weather.

Distribution and habitat

The plant generally grows on rocky slopes and foothills in the low mountains of West Texas. It has been recorded in the Trans-Pecos, Edwards Plateau and Plains regions. Regions: 7, 8, 9, 10.

Toxic agent

The toxin involved in buckwheat poisoning is a photodynamic agent. The animal absorbs this substance directly from its digestive tract (primary photosensitization) and can transfer it to calves through the milk. Once this process occurs, the animal becomes hypersensitive to sunlight.

Livestock signs

Signs of photosensitization include:

- Reddening of light-colored skin, especially thin-skinned areas and those having thin or no hair, such as the muzzle, udder and vulva
- Skin inflammation followed by swelling, blisters, fluid seepage and usually sloughing of the skin
- For dark animals, the skin is not blistered or sloughed, but it usually becomes thickened and crusted.

Integrated management strategies

Animals with primary photo-sensitization (as compared to hepatogenous photosensitization) seldom die if proper precautions are taken. As soon as clinical signs begin, place affected animals in the shade with feed and water. Painting or spraying the affected skin with methylene blue solution or some other nontoxic dye helps protect the areas from further sun exposure.

Move unaffected animals to a new pasture, free of the plant causing the photosensitization.

90

↖ Flower

Woody base ↗

Whole plant ↓

16

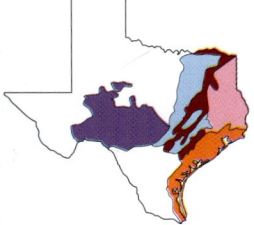

White Snakeroot
Eupatorium rugosum

White snakeroot is an erect, opposite-branching, herbaceous perennial growing to 1 to 4 feet tall. It arises from a fibrous root crown that may have short rhizomes. The slender, round stems tend to be purplish, especially when growing in the open.

The opposite leaves have three distinct veins and coarsely toothed margins. Small clusters of white flowers are produced at the ends of the branches.

Distribution and habitat

White snakeroot is most frequently found in wooded areas, but may persist after clearing.

Dense populations are found on the floors, slopes and walls of Hill Country canyons with northern exposures. These plants are found in east, southeast and north central Texas. Regions: 1, 2, 3, 4, 7.

Toxic agent

The toxic agent of white snakeroot is tremetone, which affects all animals, including humans. It is efficiently passed into the milk, which protects the lactating female but poisons the young.

Most white snakeroot poisonings in Texas occur in sheep and especially in goats, where the lethal dose is only about 0.5 percent of an animal's body weight.

Tremetone degrades as the plants dry, so plants in hay are less hazardous than the green plant.

Animals raised in pastures with the plant avoid it and are seldom poisoned. Introducing naive goats to the plant in the spring has caused death losses of more than 90 percent.

Livestock signs

Clinical signs of snakeroot poisoning, although similar, vary with the species affected. Signs in sheep and goats are predominately those of liver failure:

- Depression
- Weakness
- Head pressing
- Slight muscle tremors after exercise
- Coma
- Death in 1 to 5 days

In cattle, the signs include:

- Listlessness
- Stiff movements
- Severe muscle tremors after exercise
- Coma
- Death in 3 to 21 days

Horses show clinical signs similar to those of cattle except that there are also some signs related to disrupted cardiac function (abnormal heart rate, jugular distension and ventral edema), and death can occur in 1 to 2 days.

Integrated management strategies

Goats naive to white snakeroot should never be allowed access to it except in midwinter when there are no green plants.

Cattle and horses should not be forced to consume the plant because of poor range condition. Many cattle showing signs of tremors will recover if they are moved from contaminated pastures and allowed to remain quiet with as little disturbance as possible. Do not expose horses to the growing plants.

Leaf ↗

Flower ↘

↙ Whole plant

Snow-on-the-Mountain, Snow-on-the-Prairie
Euphorbia marginata, E. bicolor

Snow-on-the-mountain and snow-on-the-prairie are annual herbs in the spurge family. They generally grow to 1 to 3 feet tall.

The flowering stems have a peculiar construction: a whorl of four to five petal-like members, usually yellow-green, surrounding a cluster of male flowers, each consisting of a single stamen.

E. marginata leaves are long, oval and nearly hairless and come to a blunt point; the upper leaves usually have distinct white margins. *E. bicolor* leaves are similar but narrower.

Distribution and habitat

Snow-on-the-mountain is locally abundant in Central Texas. It is somewhat uncommon in the Rio Grande Plains and Trans-Pecos regions. It also grows north to Montana and Minnesota and south to Mexico. It is found most often in tight clayey soils of swales and meadows and in dry stream beds. Populations can vary greatly from year to year.

Snow-on-the-prairie is commonly found in the eastern third of Texas. Regions for *E. bicolor*: 1, 2, 3, 4; for *E. marginata*: 5, 6, 7, 8, 9, 10.

Toxic agent

The white sap of these plants has long been used to blister the skin or as an intestinal purgative. In most cases, livestock are poisoned by an acrid principle that severely irritates the mouth and gastrointestinal tract. This plant rarely causes death.

Experimental feedings of this plant in Texas have shown that 100 ounces produces severe scours and weight loss in cattle, the latter persisting for several months.

Livestock signs

Primary signs include:
- Severe irritation of mouth and gastrointestinal tract
- Diarrhea

Integrated management strategies

Most species of spurge can be grazed to a limited degree without problems. Their bitter white juice apparently makes most species unpalatable. Administer intestinal astringents if needed to relieve diarrhea and intestinal wall irritation.

Although most broadleaf herbicides control the plant easily, it is generally uneconomical to control as a primary target species.

Proper mineral supplementation, especially of phosphorus, reduces livestock losses to the plant.

↖ Flower

Flower and
seed pod ↗

Whole plant ↓

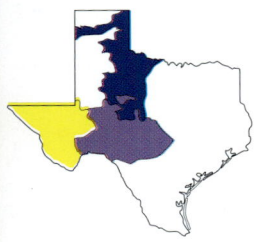

Tarbush, Blackbrush
Flourensia cernua

Tarbush is a strongly aromatic, perennial shrub and a member of the sunflower family. The plant's height may vary from 1 foot tall on dry sites to more than 6 feet in deep, overflow areas.

The leaves are alternate, smooth along the edges and oval or oblong. Flowers are solitary in the leaf axils, forming a leafy flowering stem in the fall.

Distribution and habitat

Tarbush is common on dry plains, hills and mesas from counties just east of the Pecos River in Texas, west to Arizona and south to Mexico. Regions: 7, 8, 10.

Toxic agent

Tarbush may poison sheep, goats and cattle. The toxin in tarbush is unknown. In experiments, sheep and goats were force-fed tarbush fruits, which were lethal at about 1 percent of the animal's weight. Individual susceptibility varied considerably, and the difference between toxic and lethal doses was small. In moderate amounts, the foliage was not toxic.

Livestock signs

On the range, clinical signs appear a day or less after an animal eats a toxic amount of the plant. Depending on the amount ingested, death may follow within 24 to 72 hours, or occasionally longer. Some animals recover rapidly after showing clinical signs for several days to a week.

Signs are modified somewhat by the severity of poisoning, but generally consist of:

- Loss of appetite
- Abdominal pain
- Reluctance to move
- Occasional respiratory distress

Acute cases may salivate profusely. Animals generally remain on their feet until shortly before dying without struggle. Poisonings occur mostly in January through March after the fruit has matured but before it falls.

Integrated management strategies

Tarbush is extremely unpalatable, and is grazed only if animals are starving or have severe phosphorus deficiencies. Avoid overgrazing and use proper supplemental feeding programs (including phosphorus).

Spike 20P® controls tarbush on a large-broadcast basis. For aerial or ground broadcast applications, apply Spike 20P® at 0.75 to 1 pound a.i./acre (3.75 to 5 pounds of pellets).

Follow herbicide treatments with proper stocking rates and good grazing management practices.

↖ **Leaf**

Flower ↗

Whole plant ↓

97

Perennial Broomweed, Broom Snakeweed
Gutierrezia sarothrae

Perennial broomweed or broom snakeweed is a short-lived, perennial half-shrub ranging from 6 inches to about 2 feet tall. Many unbranched, erect stems originate from a woody base and die back when the plant goes dormant.

The leaves are narrow and threadlike. The small yellow flowers are clustered at the branch tips from June to October.

Distribution and habitat

Perennial broomweed is widespread on dry ranges and deserts from California to Texas, south to Mexico and north to Idaho.

Extreme infestations reduce forage production but may not indicate overgrazed ranges because broomweed populations fluctuate naturally. However, overgrazing does accelerate the plants growth and propagation. Regions: 2, 3, 4, 5, 6, 7, 8, 9, 10.

Toxic agent

Perennial broomweed poisons cattle, sheep, goats and swine. Some believe the toxic agent is a steroidal saponin. The plant may accumulate selenium when on high selenium soil. Perennial broomweed is most toxic at earlier growth stages, usually in late winter or early spring and on sandy soils. It is relatively nontoxic growing on clay soils.

Cattle abort after eating as little as 20 pounds of fresh broomweed in 7 days. Cattle, sheep and goats have been killed by eating 10 to 20 percent of their body weight in perennial broomweed over 2 weeks.

Livestock signs

Chronic poisoning signs include:
- Abortion
- Stillbirth
- Retained placenta
- Weak offspring

Acute poisoning signs include:
- Periodic, thick nasal discharge
- Crusting and sloughing of the skin of the muzzle
- Listlessness
- Loss of appetite and weight
- Rough hair coat
- Occasionally, dark brown or reddish urine

Frequent urination with diarrhea in the early stages changes to constipation with large amounts of mucous. Pregnant cows often have periodic vulvar swellings and premature udder development.

Integrated management strategies

Do not graze gestating livestock on sandy soils during maximum perennial broomweed growth (late winter and early spring). Proper range management practices that improve range condition may help limit perennial broomweed population densities and livestock consumption.

Carefully watch pregnant cattle grazing perennial broomweed-infested pastures. If udders develop prematurely, vulvas swell or abortions occur, move the herd to a perennial broomweed-free pasture.

Chemical control may be cost-effective when populations are dense enough to reduce forage production. Herbicide treatments for perennial broomweed include aerial or ground broadcast applications of Escort® at 0.0625 ounce a.i./acre in the spring when weeds are less than 4 inches high. Tordon 22K® may be used during and after full flower stage in the fall at 0.25 to 0.5 pound a.i./acre.

To reduce reinfestation, follow chemical treatments with proper range and livestock management programs.

Flower ↗

Whole plant (spring) ↘

↙ Whole plant (fall)

99

Bitter Sneezeweed
Helenium amarum

Bitter sneezeweed is an erect, upper-branching annual, 10 to 20 inches tall with narrow leaves, alternating on the stem. The flowers are noticeable in the late spring or summer and are located at the end of each branch.

Two varieties of this plant are identical except for the flower color: one is all yellow; the other is yellow with a red-brown center. Each bloom has about eight cleft ray flowers (petals) with three lobes, often bending downward at maturity.

In some years, the lower leaves are lost, new growth occurs up the stalk and new flowers appear in the fall. The entire plant has a strong odor and is bitter to the taste.

Distribution and habitat

Bitter sneezeweed is found in all vegetational areas of Texas. The yellow variety is widespread in disturbed, sandy or loamy soil in the eastern part of the state, while the dark-centered variety is often found in calcareous soil in more western areas. Regions: 1, 2, 3, 4, 5, 6, 7, 8, 9, 10.

Toxic agent

A sesquiterpene lactone is responsible for the toxicity of bitter sneezeweed, which is greatest at time of flowering. This bitter plant is seldom consumed at a level high enough to produce clinical signs. However, it has been responsible for bitter, undrinkable milk and is suspected to be the cause of unpalatable meat from calves slaughtered off the range. The toxin is stable in plants contaminating hay.

Livestock signs

Signs of bitter sneezeweed poisoning include:
- Weakness
- Incoordination
- Vomiting
- Salivation
- Diarrhea
- Grinding of teeth

Integrated management strategies

Avoid cutting hay containing a large amount of bitter sneezeweed. Do not feed hay containing any of the plant to lactating dairy cows. Do not slaughter grass-fed cattle from a pasture that contains bitter sneezeweed.

Severe infestations may be controlled with broadleaf herbicides such as 2,4-D or Grazon P + D® at 0.5 to 1.0 pound a.i./acre in the spring with good growing conditions.

↑ Whole plant
↙ Flower
Whole plant ↘

101

101

Smallhead Sneezeweed
Helenium microcephalum

Smallhead sneezeweed is an erect, branching, annual (sometimes biennial) herb with alternate, oblong leaves that end in a point. The leaves extend down the stem to form a winged stem.

The plant is a composite (sunflower family), with many small yellow flowers growing from May to October. The disk flowers (central cone) appear somewhat spherical and are equal to or more obvious than the yellow ray flowers (flower petals).

Distribution and habitat

Smallhead sneezeweed is usually found in moist habitats around tanks and ponds and in bar ditches. A wet fall and spring usually assures a good crop of seedlings.

Common in the western half of Texas, it also may be found throughout the state except in the East Texas Piney Woods. It grows southward into Mexico. Regions: 2, 3, 4, 5, 6, 7, 8, 9, 10.

Toxic agent

The toxic agent is a sesquiterpene lactone commonly referred to as dugaldin (helenalin). Sheep are most often poisoned, although goats and cattle also may be susceptible. Horses and mules are also very sensitive to the toxin.

Smallhead sneezeweed is extremely toxic in the flowering stage.

In feeding studies, as little as 0.25 percent of an animal's body weight produced acute poisoning and death. Mature plants are more toxic than seedlings, and the winter basal rosette stage seems to lack toxicity.

Livestock signs

Sneezeweed poisoning is nicknamed "spewing sickness." Signs of illness appear within a few hours of consumption and include:

- Weakness and staggering
- Diarrhea
- Vomiting green material
- Green salivation and nasal discharge
- Bloating
- Groaning and grinding of teeth
- Sticky nonpelleted feces
- Gastroenteritis

Respiration is usually forced and fast, and the pulse, irregular and rapid. Animals may convulse immediately before death.

Postmortem examination may reveal an inflamed gastrointestinal tract with edema in the walls of the stomach, lungs and membranes surrounding the lungs and lining of the abdominal cavity. Also, hemorrhages may be evident within the heart chambers.

Integrated management strategies

In early poisoning stages, administer large doses of mineral oil or intestinal purgatives to help prevent losses. Fencing livestock away from localized infested areas during plant maturation can eliminate potential problems.

Because it is usually in localized areas, hand-pulling, mowing or burning may be effective also. The plant is highly susceptible to most broadleaf herbicides. Before plants mature, small populations may be treated with ground broadcast applications.

Scout potential habitat areas in the fall when the number of plants in the basal rosette stage can indicate the potential severity of a problem. Control the plant in fields used for green chop or hay.

Seedling ↗

Leaf →

Whole plant ↘

↙ Flower

Western Bitterweed
Hymenoxys odorata

Bitterweed is an erect, annual, composite plant growing from 3 inches to 2 feet tall. Stems are purplish near the base. Leaves are alternate and usually woolly underneath.

Bright yellow flowers bloom from April through June and occasionally in the fall. This plant has a bitter taste and a distinct odor.

Distribution and habitat

Bitterweed is common in arid areas of the southern Great Plains from southwestern Kansas and central Texas to southern California and into Mexico. It is most common where soil disturbance or overgrazing has occurred. Populations can be quite variable between years. Regions: 3, 4, 5, 6, 7, 8, 9, 10.

Toxic agent

Bitterweed is toxic to sheep and is generally unpalatable. However, starved sheep that begin eating the plant may develop a liking for it.

Cases of poisoning in cattle, horses or goats are rare. The toxic agent is a sesquiterpene lactone (hymenoxon). This material appears to accumulate, with a lethal dose consisting of 1.3 percent of an animal's weight in green plant material, whether ingested at one time or over several months. The minimum lethal dose varies considerably among individual animals, regardless of their nutritional history.

From a management perspective, it is probably best to consider bitterweed toxic at all growth stages. The plant becomes much more toxic in drought, when a lethal dose is 0.5 percent of the animal's body weight.

Livestock signs

Signs of acute bitterweed poisoning include:

- Loss of appetite
- Rumin stasis
- Depression
- Indications of abdominal pain (arched back stance)
- Bloating

Green salivary and nasal discharge is a typical sign on the range. Weight loss is the most common sign of chronic bitterweed poisoning. Clinical signs are usually not immediate; they may appear a month or more after the plant is first eaten.

Integrated management strategies

There is no medical treatment for bitterweed poisoning. However, recent evidence indicates that activated charcoal may alleviate signs.

Move poisoned animals to bitterweed-free pastures or feed them alfalfa or a suitable diet in a feedlot for 10 days before returning them to rangeland.

To prevent bitterweed losses, use a proper supplemental feeding program and high numbers of sheep in infested pastures for short periods. Removal of the sheep from pastures when bitterweed signs start requires close observation.

Large, dense populations of the plant may be treated with herbicide by aerial application. Treat localized populations of bitterweed with targeted ground applications.

Refer to Extension publication B-1466, *Chemical Weed and* *Brush Control Suggestions for Rangeland,* for specific herbicides and rates.

Flower ↗

Seedling →

↓ Whole plant

105

Rayless Goldenrod, Jimmyweed
Isocoma wrightii

Rayless goldenrod is a low-growing half-shrub with erect stems arising from a woody crown to a height of 2 to 4 feet. The leaves are sticky, narrow, alternate and may be even or slightly toothed along the margins. The stems bear flat-topped clusters of yellow flowers from June through October.

Distribution and habitat

Rayless goldenrod is often found on dry rangelands, especially in river valleys, along drainage areas and irrigation canals, and on gypsiferous soil outcrops.

It is a local problem in the Pecos Valley drainage in southeastern New Mexico and western Texas. It usually grows at 2,000 to 6,000 feet elevation and is found from southern Colorado into Texas, Mexico, New Mexico and Arizona. Regions: 8, 9, 10.

Toxic agent

Goldenrod can poison all species of livestock. The toxic agent is tremetone. The poison accumulates in the animal and is present in green and dry leaves, making the plant toxic year-round.

The toxin in rayless goldenrod can be passed through milk. It is common for poisoning signs to appear in suckling young, but not their mothers. Humans have been poisoned by consuming milk from affected cattle. Most poisoning cases occur in late fall or early winter, but can occur year-round.

A lethal dose generally consists of 1.0 to 1.5 percent of the animal's weight, consumed over 2 to 3 weeks.

Livestock signs

In cattle, this plant produces clinical signs often referred to as the trembles. Muscular trembling is particularly noticeable about the nose, hips and over the shoulders. Trembling is more pronounced after exercise.

Stiffness and weakness are most pronounced in the forelegs. In later stages, the animal lies down and becomes unable to rise.

Other signs may include:
- Constipation
- Vomiting
- Quickened and labored breathing
- Almost continuous dribbling of urine

Shortly before death, the animal breathes with a prolonged inhalation followed by a pause and then a short and somewhat forcible expiration. Postmortem findings in cattle include:
- Congestion of the abomasum and intestine
- Pale liver
- Distended gall bladder

Integrated management strategies

Keep animals away from areas severely infested with rayless goldenrod.

There is no specific treatment for poisoned animals. Remove them from the area of poisoning and give them good-quality hay and water. Purgatives and laxative feeds may aid recovery. Orally administering activated charcoal at 1 gram per kilogram of body weight may be helpful. Take calves and lambs off poisoned mothers. Discard all milk from affected females.

Chemical control is achieved through aerial or ground broadcast methods in the fall. Good results have come from applying 0.45 ounces a.i./acre of Escort® or 0.5 pound a.i./acre of Tordon 22K®.

Treat individual plants with a 1.0 percent v/v solution of Tordon 22K®.

Flower ↗

Whole plant (fall) →

Whole plant ↓

Narrowleaf Sumpweed
Iva angustifolia

Narrowleaf sumpweed is an inconspicuous member of the sunflower family. It is an unpalatable annual, 20 to 45 inches tall and quite branched. The stems of young plants are often purplish.

The opposite, lance-shaped leaves are up to 5 inches long and less than 0.5 inch wide, entire or only slightly serrate (notched), and are reduced to bracts in the head-bearing region. The small green flowers have no petals and are located in elongated heads at the end of each branch.

Distribution and habitat

These plants are located in all vegetational areas of Texas except those of the Panhandle.

In areas of higher rainfall, narrowleaf sumpweed tends to be located in elevated, well-drained areas. In regions of low rainfall, it grows in drainage ditches, along creek beds and subirrigated areas. Regions: 1, 2, 3, 4, 5, 6, 7, 10.

Toxic agent

The toxin in narrowleaf sumpweed is unknown. This plant is often very abundant and is unavoidably grazed heavily. Consumption has been associated with abortions at 4 to 8 months of gestation when cattle ate large amounts of the young plants in the two- to eight-leaf stage of growth.

In experiments, rabbits fed the plant as 50 percent of their diet either gave birth prematurely or had stillborn or weak pups that died by 3 days old.

Livestock signs

Cattle consuming a large amount of young narrowleaf sumpweed in mid-gestation can show the following signs:

- Premature mammary development
- Dripping milk
- Abortion

Affected cows breed back normally.

Integrated management strategies

Cows in mid-gestation should not be placed in dry pastures that contain subirrigated areas heavily infested with narrowleaf sumpweed if little forage is available in other areas. Time of breeding can be altered so that plants are large enough for cows to avoid them during mid-gestation.

Severe infestations may be controlled with broadleaf herbicides such as 2,4-D or Grazon P + D® at 0.5 to 1.0 pound a.i./acre in the spring with good growing conditions.

108

↖ Seedling

Flower ↗

Whole plant ↓

Berlandier Nettle Spurge
Jatropha cathartica

Berlandier nettle spurge is a perennial herb that grows from enlarged, tuberlike woody roots up to 10 inches thick. The hairless stems, 4 to 10 inches tall, are branched and spreading.

Its palm-shaped leaves are up to 4 inches long and very deeply lobed five to seven times. Showy red flowers up to 0.5 inch in diameter are arrayed in loose clusters at the ends of the stems. The fruit is a three-lobed capsule containing three seeds.

Distribution and habitat

In Texas, these plants are limited to the Rio Grande Plains. They can be found scattered among the brush growing on clay soil. Regions: 2, 6.

Toxic agent

The toxic agent or agents of nettle spurge are not known. Very little research has been conducted on this plant. The seeds of a Central American species of *Jatropha* cause a similar disease. Those seeds contain phorbol esters, a trypsin inhibitor, a lectin and phytate.

Intoxication has been recognized in humans, chickens, goats and calves. Both cattle and penned deer have been poisoned by *Jatropha cathartica* in Texas.

The tubers allow nettle spurge to respond rapidly after rain during drought, so it is available for consumption before there is much growth of other plants.

Livestock signs

Signs of poisoning are associated with the gastrointestinal system and include:
- Weakness
- Diarrhea (at times with dark blood)
- Death

Integrated management strategies

Animals should not be forced to consume this plant. It is unpalatable, and livestock will not consume it if adequate forage is available.

Mechanically remove these plants from deer pens or any other dry-lot situation.

↖ Flower
and root

Flower and
pod ↗

Whole plant ↓

Leatherstem
Jatropha dioica

Leatherstem is an erect, perennial, shrubby plant with simple or somewhat lobed leaves. It is a member of the spurge family (Euphorbiaceae). The stems are quite flexible, thus the name leatherstem.

Distribution and habitat

Leatherstem grows on gravelly bluffs, hillsides and ravine slopes of the South Texas Plains, the Edwards Plateau and Trans-Pecos areas. It also grows south into Mexico. Regions: 2, 6, 7, 10.

Toxic agent

Leatherstem is poisonous to sheep and goats. The major toxic agents are phorbol esters, which severely irritate the stomach lining.

A limited number of feeding experiments by the Texas Agricultural Experiment Station found that leatherstem was toxic but not fatal to a sheep fed 2 percent of its body weight of green leaves. Seeds were fatal to a lamb at 3 percent of its body weight in three doses over 12 days.

A goat fed 3.7 percent of its body weight of leatherstem leaves developed progressive anemia and died.

Livestock signs

Signs for leatherstem include:
• Severe gastric inflammation
• Abdominal pain
• Vomiting
• Diarrhea

Integrated management strategies

Leatherstem is relatively unpalatable to livestock. Grazing management practices that improve or maintain good quality range condition and avoid overgrazing reduce the incidence of poisoning from leatherstem.

Proper mineral supplementation programs, especially providing phosphorus, also reduce livestock losses to the plant.

↖ Leaf

Fruit and
whole plant ↗

Whole plant ↓

113

Hairy Caltrop, Warty Caltrop
Kallstroemia hirsutissima, K. parviflora

Hairy caltrop is a much-branched annual with long, prostrate limbs emanating from a central root. Its growth habit and general appearance are very similar to goathead (puncturevine). Both are members of the caltrop family.

The fruit of hairy caltrop is beaked, and at maturity it breaks up into eight to 12 one-seeded nutlets. Stems and leaves are conspicuously hairy; the beak of the fruit is no more than 3 millimeters long.

A close relative to hairy caltrop is warty caltrop *(K. parviflora)*. The plants are quite similar, with only slight differences in the fruits.

Distribution and habitat

Caltrops are typically found in old fields, disturbed areas and heavily grazed pastures. Both species are distributed widely across Texas.

Hairy caltrop is more common in the western part of the state; warty caltrop is more common in the east. Hairy caltrop is also found in southeastern Arizona, southern New Mexico and Mexico.

Warty caltrop can be found from Texas to California, north through Oklahoma, Kansas and Missouri to Illinois and Mississippi. Regions: 2, 3, 4, 5, 6, 7, 8, 10.

Toxic agent

These plants are toxic to cattle, sheep, goats and rabbits. The toxic agent is unknown, although it is believed that an animal must eat about one-third of its weight in caltrop to be poisoned.

Livestock signs

The first proven cases of cattle losses to hairy caltrop in the Trans-Pecos region of Texas were documented in 1944. The progression of signs in cattle and goats include:

• Weakness in the hind legs with knuckling of the fetlock joint

• Posterior paralysis

• Frequently, convulsions before death

Sheep losses seem to be sporadic, with signs of illness similar to those in cattle, but sheep walk on their front knees before convulsions develop.

Integrated management strategies

No medicinal treatment is available to counteract caltrop poisoning. As soon as poisoning is suspected, move animals to pastures free of caltrop. Do not handle or drive affected animals excessively.

Problems generally occur when pastures are overgrazed and no other forage is available.

Proper range management practices can alleviate most problems associated with this plant.

In general, do not graze areas where most of the forage is hairy or warty caltrop.

↖ Leaf

Flower ↗

Whole plant ↓

Coyotillo
Karwinskia humboldtiana

Coyotillo is a spineless woody shrub or small tree of the buckthorn family. Plants on rangelands are generally 1 to 5 feet tall, with crown diameters up to 10 feet.

The most distinctive feature of this plant is the leaves, which are opposite and have prominent veins ending at the leaf's untoothed margin. Small greenish flowers, produced in the leaf axils, have a five-lobed calyx, five petals, five stamens and a compound pistil.

The ovary develops into a brownish-black, oval-shaped fruit or berry. The berries mature in late summer and fall.

Distribution and habitat

Coyotillo grows along arroyos and canyons and on gravelly hills and ridges. It is found from southwest Texas into Mexico and southern California. Regions: 2, 6, 7, 10.

Toxic agent

Coyotillo contains polyphenolic compounds thought to be responsible for its toxicity. This plant is toxic to cattle, sheep, goats, hogs and fowl; cattle are the most susceptible. Native Americans knew that the berries from this plant produced paralysis in humans.

Consumption of 0.05 to 0.3 percent of an animal's weight in seeds will produce poisoning. The leaves are much less toxic, requiring consumption of 15 to 21 percent of the animal's weight for poisoning. Seeds and leaves produce two different poisoning syndromes.

There is usually a lag period of days or weeks between seed ingestion and the appearance of livestock signs. Problems from browsing this shrub occur most often in fall and winter.

Livestock signs

Signs of seed ingestion include:
- Exaggerated high stepping
- Jumping or moving backward
- Weakness and incoordination of hind legs
- Dragging motion while walking
- Complete prostration
- Death

Signs of foliage ingestion include:
- Loss of condition
- Wasting
- Nausea
- Progressive weakness
- Death

Integrated management strategies

No effective medical treatment is known for coyotillo poisoning. Few animals paralyzed from eating the seed recover under range conditions.

Do not turn animals (especially goats) unfamiliar with coyotillo into infested areas when this shrub is in fruit. Proper stocking and good supplemental feed and mineral programs help reduce the incidence of poisoning.

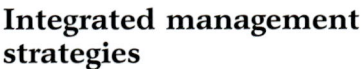

Leaf and berries↗

Whole plant ↓

Kochia
Kochia scoparia

Kochia is an annual forb growing from 2 to 5 feet tall beginning in late spring. Stems are erect, much-branched and leafy. Leaves, arranged alternately on the stems, turn bright red with age.

Interest in the plant as a feed source has been high because it can produce the same amount of forage per acre as alfalfa with half the water; hence, the plant is often called "poor man's alfalfa."

Distribution and habitat

An annual plant originating in Eurasia, kochia was introduced into America in the early 1900s. Its appearance was first recorded in Texas during the late 1940s.

Found throughout most of the United States, it occurs locally on disturbed sites and in old fields. Regions: 2, 7, 8, 9, 10.

Toxic agent

The plant contains a number of agents that could contribute to its toxicity: oxalates, alkaloids, saponins, nitrate and sulfate. Oxalates bind calcium, and a rapid drop in blood calcium levels can cause sudden death.

The saponins may damage the liver in some animals, causing photosensitization. Other cases may develop polioencephalomalacia, possibly because of the high sulfate.

Livestock signs

Several potential syndromes from kochia poisonings include (see Animal Conditions for specific signs):

- Polioencephalomalacia ("polio")
- Oxalate poisoning
- Photosensitization
- Liver damage
- Nitrate poisoning

Integrated management strategies

Two treatments have been used with varying degrees of success to counteract oxalate poisoning from kochia: injecting calcium gluconate intravenously; and feeding dicalcium phosphate free-choice at 25 percent of the usual salt ration. The latter treatment is often used as a preventive measure when cattle or sheep are grazing pastures heavily infested with kochia.

Use caution when cattle grazing kochia are allowed access to water high in salt. Some ranchers lowered toxicity problems by feeding free-choice hay straw to their animals. Treat polio cases with vitamin B_1 (thiamine).

In general, do not graze cattle on kochia for more than about 60 days. Despite its drawbacks, kochia can sometimes be considered a valuable forage.

↖ **Seedling**

Leaf ↗

Whole plant ↓

119

Largeleaf Lantana
Lantana spp.

Lantana is a branching shrub up to 6 feet tall, with spreading, ascending branches usually having a few small prickles. New growth has square stems. The branches are opposite and arise from leaf axils. The oval leaves are rough and have serrate edges.

The many-flowered heads are on long stems usually arising from the axils of the leaves, and are often of two colors. There are pink and white, yellow and orange and orange and red varieties. Some of the newer ornamental varieties have single-colored flowers. The clustered, round fruits are about ⅛ inch in diameter and are black when ripe.

Distribution and habitat

This shrub was widely planted as an ornamental as the state was settled. It is still a common shrub and has escaped in many areas. It often grows under brush and along fences, where birds apparently deposited the seeds, in all areas except deep East Texas, the western panhandle and the Trans-Pecos area. Regions: 2, 3, 4, 5, 6, 7 and 8.

Toxic agent

Triterpenes lantadene A and lantadene B from largeleaf lantana are responsible for the toxicity of the plant, which affects cattle, sheep, goats, horses, dogs and humans.

The degree of liver injury produced by the plant directly reflects the amount of lantana ingested. Low levels give slight liver damage, producing increases in liver enzymes present in the serum. Higher amounts result in cholestasis and microscopic changes in the liver. Very high doses result in widespread liver necrosis or death of liver cells.

Livestock signs

Although horses usually do not exhibit signs of photosensitization, the other clinical signs are similar in all species and include:

- Sluggishness
- Weakness
- Bloody diarrhea
- Jaundice (yellow whites of the eyes, yellow skin; yellow fat and liver after death)
- Secondary photosensitization

Integrated management strategies

Livestock should not be forced to consume lantana. Good range management with adequate palatable forage will prevent excess consumption.

Allow poisoned animals to remain in the shade and give them sun-bleached hay, feed and water.

↖ **Flower and fruit**

Leaf ↗

Whole plant ↓

121

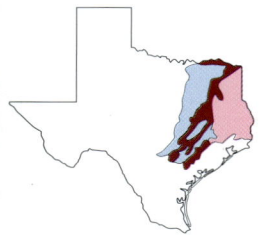

Singletary Pea
Lathyrus hirsutus

Singletary pea is a legume with winged stems that grow from 10 to 40 inches long. Its compound leaves have two long, narrow leaflets up to 3 inches long and terminate in a branched tendril.

The small, pea-like flowers are red to bluish. Distinctive pods, 1 to 1.5 inches long, are covered with hair attached to small, raised bumps. Each pod contains four to 10 mottled, round seeds.

Distribution and habitat

Historically, singletary pea was planted as a cover crop and a cool-season forage, often mixed with small grains. It has escaped cultivation and is found in north central and northeastern Texas. It is common along roadsides and in pastures where it has been allowed to go to seed. Regions: 1, 3, 4.

Toxic agent

The vegetation of singletary pea is not toxic and is highly nutritious, but the seeds contain toxic amino acids. Lathyrism, the neurological syndrome most often produced by chronic consumption of the seeds, can affect all species including humans, but horses are the most sensitive.

Horses are usually affected by hay containing intact pods with seeds. Bovine cases usually result from grazing pastures with many mature plants.

Livestock signs

Horses with lathyrism demonstrate these signs:

• Incoordination of rear legs
• Unusual stance with rear legs too far forward
• Exaggerated stepping of rear legs
• Paralysis of rear legs

Cattle with lathyrism show:

• Reluctance to stand
• Incoordination of rear legs
• Inability to rise

Chronic consumption of seeds of other *Lathyrus* species result in skeletal deformities in growing animals. Calves born to cows that have consumed seeds of singletary pea for several months during gestation may have crooked legs and a curved spine.

Integrated management strategies

Remove horses from hay containing singletary pea seeds as soon as signs develop; most will recover within 4 to 6 weeks. The condition becomes irreversible if the animals continue to eat the seeds. Cattle removed from the pasture recover in a few days.

Pastures with singletary pea should be grazed before seed production. Do not feed hay containing seedpods to horses.

↖ **Leaf**

Flower ↗

Pod ↗

↓ **Whole plant**

Gordon Bladderpod
Lesquerella gordonii

Gordon bladderpod is a multi-stemmed, annual forb in the mustard family. The root system is relatively weak, which is often a characteristic of annual plants. This can be helpful when trying to distinguish Gordon bladderpod from other bladderpods, which are perennial and have heavier, deeper root systems.

The leaves of Gordon bladderpod are sparse, small, lance-shaped and covered with silvery-gray hair. The small, four-petaled flowers are bright yellow and fade to reddish as they mature. The hollow seedpods are spherical, about ⅛ inch in diameter.

Distribution and habitat

These plants are found in gravelly or sandy soil in pastures, in open fields, on hillsides and along roadsides in the western half of Texas. Regions: 5, 6, 7, 8, 9, 10.

Toxic agent

The toxic agent of bladderpod is not known. Although this plant has been associated with "stocking up" (swelling of the lower legs) in horses, poisoning has not been confirmed by experimental trial. Feeding hoary alyssum *(Berteroa incana)*, a mustard of the northern United States, reproduced the clinical signs.

Problems are seen in Texas when a wet winter and spring follow a drought, and horses on pasture are exposed to many of these plants in the flower and seedpod stage.

Livestock signs

Only horses are affected and they may show the following clinical signs:

* High temperature
* Depression
* Edema of the lower body
* Swollen legs
* Founder

Integrated management strategies

In spring, do not place horses in pastures "yellow" with Gordon bladderpod.

Horses usually respond favorably if they are removed from the contaminated pastures in the early stage of disease. These plants must constitute a large portion of the diet to induce poisoning.

Good pasture management along with adequate desirable forage will prevent poisoning.

↖ Pod

Flower ↗

Whole plant ↓

Berlandier
Lobelia
Lobelia berlandieri

Lobelia is a cool-season annual reaching up to 20 inches tall, although most plants are less than 12 inches tall. Most of its leaves are basal, hairless, oval and up to 2 inches long on short stems. Leaves on the stalks are lance-shaped and much smaller.

Berlandier lobelia has from one to 20 thin, ascending branches, each bearing a loose stalk of small, bright, purplish-blue flowers with white eyes.

Distribution and habitat

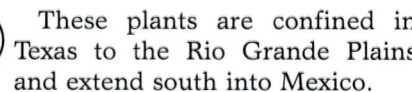

These plants are confined in Texas to the Rio Grande Plains and extend south into Mexico.

They are abundant on disturbed rocky, sandy or clay soils in years with adequate fall and winter rainfall. Regions: 2, 6.

Toxic agent

Berlandier lobelia contains nicotine alkaloids that are responsible for its toxicity. The plants are short and inconspicuous until the flowering stalk stage of growth. These plants are palatable and are readily consumed.

Above-average rainfall during the fall and winter almost always results in lobelia poisoning in cattle on the plains south of Kingsville TX. Nelgi antelope have also been poisoned in the same area.

Livestock signs

Depending on the dose, the nicotine alkaloids are central nervous system stimulants or depressants and result in the following signs:

- Excitability (early, and not usually observed in brush pastures)
- Depression
- Down animals that refuse food and water

Animals become so severely depressed that they lie down and die from water deprivation and exposure.

Integrated management strategies

Depressed and down animals often recover if they are moved to the shade and given adequate nutrition (feed and water) by tube. This treatment may have to be repeated daily for up to 2 weeks.

Prevention of poisoning requires moving cattle to pastures not containing a significant amount of the plant in the flower and seed stages.

Severe infestations may be controlled with broadleaf herbicides such as 2,4-D or Grazon P + D® at 0.5 to 1.0 pounds a.i./acre in the spring with good growing conditions.

↖ Leaf

Whole plant ↗

Flower ↓

Chinaberry
Melia azedarach

Chinaberry is a tree that grows up to 50 feet tall with a broad, spreading, rounded crown. The large leaves of up to 15 inches long have leaflets of up to 3 inches long.

Loose clusters of purplish, fragrant flowers give rise to hanging clumps of smooth single-seeded fruits about 0.5 to 0.75 inch in diameter. The clusters of fruit ripen to yellow in the fall and often persist on the tree through the winter.

Distribution and habitat

These introduced trees were widely planted as ornamentals in the eastern two-thirds of Texas. Many have escaped and may be found in thickets, floodplain woods and borders of woods. Regions 1, 2, 3, 4, 5, 6, 7, 8.

Toxic agent

Meliatoxins A1, A2 and A3 are responsible for the toxicity of these plants. They are found in highest concentration in the fruit, but the bark, leaves and flowers are also poisonous.

Many species—including cattle, sheep, goats, pigs, dogs, rats, rabbits, guinea pigs, poultry and humans— have been poisoned by chinaberry.

Pigs and dogs are poisoned most frequently, usually by ingesting fallen fruits. They show clinical signs within 2 to 4 hours of consumption.

Livestock signs

The clinical signs are related to the gastrointestinal and/or nervous system and include:
- Anorexia
- Vomiting
- Diarrhea
- Constipation
- Excitation
- Incoordination
- Depression
- Convulsions

Animals surviving for 24 hours have a good chance of recovery. Rarely do clinical signs last longer than 48 hours. Most cases result from animals consuming the fruit from the ground.

Because this intoxication is rapid, the hard, ribbed pit from the fruit is easily identifiable from the stomach contents of dead animals.

Integrated management strategies

Remove chinaberry trees from the area of pens used for swine. Other animals should not be forced to consume the bitter fruit.

↖ Berries
and leaf

Flower ↗

Whole plant ↓

129

Sweetclover
Melilotus spp.

There are three species of sweetclover in Texas. They are biennial or annual with erect, branching stalks 1 to 8 feet high. The alternate leaves have three egg-shaped leaflets with toothed margins on the tip.

Small white or yellow flowers are presented in upright terminal spikes. Crushed plants have a characteristic sickly sweet, sharp odor.

Distribution and habitat

Sweetclovers are grown throughout the mid- and south-western United States as high-yielding, high-quality cultivated forage crops. Most areas in the eastern two-thirds of Texas can grow sweetclover without irrigation. Wild plants typically grow in areas that receive extra moisture such as draws, roadsides, and stock tanks. Regions: 1, 2, 3, 4, 5, 6, 7, 8, 9, 10.

Toxic agent

Sweetclover contains coumarin, which may be converted to the toxin dicoumarol by bacteria or fungi in damaged plants, moldy hay and spoiled silage. The popularity of round-baled hay has almost eliminated the use of sweetclover, as a moldy thatch covers most bales.

Dicoumarol disrupts normal blood clotting processes, causing internal and external hemorrhaging. Hay containing 0.002 to 0.003 percent dicoumarol may poison cattle. Feeding trials with sheep have shown them to be about twice as tolerant.

Livestock signs

Clinical signs of poisoning vary depending upon the dicoumarol concentrations and the length of exposure. Signs can appear suddenly or after several months and include:

- Stiffness and lameness
- Soft swelling under the skin
- Nosebleeds
- Anemia
- Convulsions
- Sudden death

In several instances when cows have eaten sweetclover hay, the cows and/or their newborn calves have bled to death soon after parturation.

Integrated management strategies

Forage containing sweetclover should be baled only in small square bales after it is well-cured, and should not be fed if it is the least bit moldy. Chemical testing for dicoumarol in hay or silage is impractical because of difficulty in sampling.

Treat poisoned animals with vitamin K and eliminate sweetclover from their diet.

↖ **Flower** ↗

Leaf ↗

Whole plant ↓

131

Common Oleander
Nerium oleander

Oleander is an introduced, evergreen, ornamental shrub or tree (depending on pruning) 15 to 25 feet tall. Leaves are entire and leathery, essentially hairless, up to 12 inches long and 1.5 inches across. They are positioned opposite or in whorls of three or four. Each has a prominent midrib with secondary veins parallel to each other extending to the leaf margin.

The variously colored, showy, odorless flowers are produced in clusters at the ends of the branches.

Distribution and habitat

Oleander is a native of the Mediterranean region and is widely planted in the southern United States. It is often found as an ornamental in eastern, central and southern Texas. It is not cold hardy and often sustains top-kill in the winter. Although it is naturalized in Texas, there are few, if any, escaped populations. Regions: 1, 2, 3, 4, 5, 6, 7,10.

Toxic agent

Oleandrin, a cardiac glycoside, is the most prominent toxin in oleander, which is probably the most toxic plant in Texas. As little as 0.005 percent of an animal's body weight of dry leaves may be lethal—as few as 10 to 20 medium-sized leaves may kill an adult horse.

It is toxic to all animal species, and many livestock and pets are poisoned, usually because they consumed oleander clippings or dead leaves.

The green leaves of the growing shrubs are bitter and are therefore seldom eaten. The wilted clippings and dead leaves remain toxic, are palatable and are readily consumed. Compost containing oleander leaves has also been incriminated in poisoning.

Livestock signs

Consumption of this highly toxic plant causes cardiac failure. Signs in poisoned animals develop within 4 hours and can include:

- Sudden death (no observed clinical signs)
- Colic
- Weakness
- Lack of rumen muscle tone
- Salivation
- Very fast or slow heart rate

Integrated management strategies

Prevention of oleander poisoning is easy and absolute: Do not plant any on your property.

Remove plants that are present, and do not allow animals access to removed plants or clippings.

↖ **Leaf**

Flower ↗

Whole plants ↓

133

Tree Tobacco
Nicotiana glauca

Tree tobacco is a loose-branching, small, evergreen tree or shrub between 6 and 18 feet tall.

The large, lance-shaped, smooth leaves appear on short stalks opposite each other low on the branches. The upper leaves lack stalks and lie in a somewhat upward angle against the branch.

White, yellow, greenish or red, trumpet-shaped flowers open at night and are 0.5 to 1.5 inches long.

Distribution and habitat

Tree tobacco is a native of South America, but is naturalized along streams and roadsides in low elevation, semi-arid areas of the southwestern United States. In the Big Bend area of Texas, it occurs frequently along the Rio Grande and smaller streams and in southern Texas as well. Tree tobacco is also used as an ornamental. Regions: 2, 6, 7, 10.

Toxic agent

The tobaccos contain the toxic alkaloids nicotine and anabasine, and all parts of the evergreen plant are toxic year-round.

In Texas, cattle and horses are most frequently poisoned. The toxic dose for horses is 0.5 percent of body weight in leaves, while poisonings in cattle are reported at 2 percent of the body weight. Swine, sheep and probably goats are also susceptible.

Livestock signs

Signs appear shortly after consumption of acutely toxic amounts. They may include:

- Stumbling
- Muscle trembling
- Salivation
- Frequent urination
- Vomiting
- Diarrhea
- Death due to respiratory paralysis

Birth defects, crooked legs and/or cleft palate may occur in young when the mother consumed a low level during the first trimester of pregnancy.

Integrated management strategies

The plant is generally unpalatable to livestock. Maintaining good range condition and an adequate forage supply is the best way to avoid problems.

Activated charcoal may be of some use in treating acutely poisoned animals.

↖ Leaf

Flower ↗

Whole plant ↓

Desert Tobacco
Nicotiana trigonophylla

Desert tobacco is a slender, erect plant that grows to about 1 to 3 feet tall. Its stems are covered with sticky hairs.

The oval to lance-shaped leaves are 2 to 5 inches long and covered with soft, short hairs. They grow opposite each other on the stem, attached directly at the base.

The five-lobed, greenish-white or yellowish, trumpet-shaped flowers are about 0.5 to 1.5 inches long. This day-blooming plant flowers year-round.

Distribution and habitat

The plant is often found in gravelly-sandy draws and arroyos. Regions: 2, 6, 7, 8, 9, 10.

Toxic agent

All parts of desert tobacco contain the alkaloids nicotine and anabasine as toxic principles. In Texas, poisoning by this plant is rare, but cattle and horses are the species usually involved. Swine, sheep and probably goats are also susceptible.

The toxic dose for horses is 0.5 percent of body weight in leaves; for cattle, it is reported as 2 percent of body weight. Lower plant consumption can produce birth defects when the plant is eaten during the first trimester of pregnancy.

Livestock signs

These alkaloids affect the nervous and muscular systems, and the clinical signs may include:
- Incoordination
- Muscle trembling
- Salivation
- Frequent urination
- Colic
- Vomiting
- Diarrhea
- Death from respiratory paralysis
- Birth defects: crooked legs and/or cleft palate

Integrated management strategies

The plant is generally unpalatable to livestock and consumption must be forced. There may be some benefit to treating acutely poisoned animals with activated charcoal.

Maintaining good range condition and an adequate forage supply is the best way to avoid problems.

Flower ↗

Seedling ↘

← Whole plant

137

Sacahuista
Nolina texana

Sacahuista is a perennial in the lily family. The plant forms a large, distinctive clump of many fibrous, narrow leaves up to 5 feet long.

The stems are woody and mostly buried. With adequate rainfall, the plant gives rise in the spring to several stems bearing many clustered flowers. The flower stalks usually are not apparent until the plant is in full bloom.

Distribution and habitat

Sacahuista is usually found on rocky range sites and mountain foothills from 3,000 to 7,000 feet in elevation. It is found in western Texas, Arizona, New Mexico and Mexico. Regions: 5, 6, 7, 8, 10.

Toxic agent

The flower buds, blooms and fruit contain saponins toxic to the liver. On the range, sheep, goats and cattle eat these plant parts avidly. Although livestock, particularly cattle, eat the foliage to some extent, it does not poison them.

Feeding experiments have found that a minimum toxic dose for sheep is about 1.0 percent of the animal's weight in buds or blooms. Ingesting sacahuista fruit, blooms or buds causes severe liver damage.

The minimum toxic dose is very close to the minimum lethal dose. Under range conditions, almost all animals developing signs will eventually die. Goats seem to be more susceptible to sacahuista than sheep.

Livestock signs

Sacahuista produces signs of:
• Generalized jaundice
• Loss of appetite
• Liver and kidney damage
• Photosensitization
• Progressive weakness

Dermatitis with itching may occur in early stages of photosensitization.

Integrated management strategies

Only the fruit, blooms and buds of the plant are toxic. Because these parts are available in the spring, restrict sheep and goats to pastures with the least density of sacahuista and graze cattle in the more dense pastures during this time.

The severity of loss correlates with abundance of bloom, which varies greatly from year to year. Heavy blooms occur only once every 5 or 6 years on average.

Individual plants may be controlled with 4 ounces of Spike 20P® pellets per plant or mechanical grubbing. A few pastures might be made sacahuista-free and could be grazed when the plants are flowering in other pastures.

↖ Flower

Fruit ↗

Whole plant ↓

Desert Spike
Oligomeris linifolia

Desert spike is a succulent, erect, annual cool-season plant reaching 14 to 15 inches tall.

The plant branches from the base. Its numerous leaves are linear shaped and grow to 1 inch long.

The species is the only plant of the mignonette family found in Texas.

Distribution and habitat

This plant is generally found on salt and clay flats, at the feet of boulders and on gravel bars along streams in the Rio Grande Valley and Trans-Pecos regions of Texas. Desert spike grows from Texas to California and south into northern Mexico. Regions: 6, 7, 10.

Toxic agent

The toxic agent involved in desert spike poisoning is unknown. Only cattle are known to have been poisoned by this plant in Texas.

Feeding trials have shown that seedlings and mature plants are toxic to calves fed 2.5 percent of their body weight of desert spike per day for 2 to 12 days.

Livestock signs

Cattle poisoned by desert spike exhibit central nervous system signs such as:

- Nervousness
- Salivation
- Trembling and weakness
- Apparent delirium
- Rapid breathing
- Collapse
- Prolonged coma accompanied by marked trembling

Signs may differ somewhat, and some poisoned cattle have appeared blind while not actually being blind. Desert spike poisonings are easily misdiagnosed as lead poisoning.

Integrated management strategies

Cattle find desert spike unpalatable. To reduce the incidence of livestock poisoning from this plant, manage grazing so animals have enough desirable forage available.

Proper mineral supplemental feeding programs, especially those providing phosphorus, may also help reduce losses to desert spike.

Seedhead ↑

Whole plant ↓

141

Lambert Crazyweed, Loco
Oxytropis lambertii

Lambert crazyweed is a perennial legume that often forms colonies by short rhizomes, with larger plants forming mounds. Moisture and temperature greatly influence the abundance of vegetative growth, and the plants sometimes die completely back to thick root crowns.

Leaves arise directly from the root crown and are up to 8 inches long, with nine to 19 leaflets. The leaflets are hairy, and each silky hair is attached at its center with both ends free.

Pink-purple to almost white flowers are displayed in terminal spikes on leafless stalks. The pods are sessile, oblong, beaked, silky and curved backward or inward.

Distribution and habitat

The plant grows primarily on well-drained sandy or gravelly soils and often on rocky knolls in north central Texas, in the Panhandle and on the High Plains. It extends into similar regions of New Mexico, Oklahoma, Colorado and Kansas. Regions: 4, 5, 7 and 8.

Toxic agent

The toxic agent is the alkaloid swainsonine, which causes an enzyme dysfunction resulting in damage to the brain, liver, digestive organs, placenta, ovaries and testes. Cell damage is reversible except in the brain.

The plant is toxic to cattle, sheep, goats and particularly horses. Horses may display signs after consuming 30 percent of their body weight of the plant, and the lethal dose is around 75 percent of their body weight.

Cattle and sheep display signs of locoism after eating about 90 percent of their body weight over a 2-month period, and lethal doses in ruminants are usually around 200 to 350 percent of their body weight over several months.

Livestock signs

Signs result from the involvement of sensory and motor functions. In cattle, general signs include:

- Carrying the head a little lower than normal
- Vacant stare
- Trembling of the head
- Difficulty or inability to eat and drink
- Infertility or subfertility in males and females
- Abortion, or deformed or weak offspring

Swainsonine is passed into the milk, leading to the unthriftiness of some suckling calves. Poisoned horses are listless, but become excessively excited to the point of stumbling when stimulated suddenly. Horses with chronic locoism rarely recover and are permanently dangerous to ride.

Integrated management strategies

Livestock imported from areas where loco does not grow are the most susceptible to poisoning. Native animals generally avoid locoweed when good quality forage is available.

Maintaining good range condition and a sound supplemental protein and mineral feeding program is the best prevention against locoism. Individuals observed eating loco should be removed to a locoweed-free pasture.

Grazon P+D® can be applied to individual plants or broadcast to control plants dominating a particular area. Treated plants become much more toxic and palatable until they are completely dry.

Seed pods ↗

Flower ↘

↙ Whole plant

143

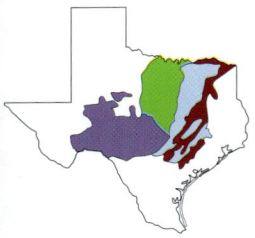

Kleingrass
Panicum coloratum

Kleingrass is a tufted perennial bunch grass with stems arising from firm, often knotty bases. It grows from 20 to 50 inches tall.

The plant gives rise to an open-flowering head with rounded seeds. The species is a warm-season grass that can provide good grazing for cattle.

Distribution and habitat

Not native to Texas, the grass was introduced from Africa in the 1950s. After more than 10 years of research, Kleingrass 75 was released as the most desirable variety, through joint efforts of the Texas Agricultural Experiment Station and the U.S. Soil Conservation Service. In the past 45 years, hundreds of thousands of acres in Texas have been planted to a monoculture of kleingrass. Regions: 3, 4, 5, 7.

Toxic agent

If managed properly, kleingrass provides abundant good-quality forage for cattle. However, saponins in the grass cause liver damage in horses, sheep and goats, with accompanying photosensitization in small ruminants. Cattle appear to be unaffected.

Green growth after moisture or grazing is reported to be more toxic than old or dormant growth.

Livestock signs

Poisoned sheep and goats exhibit typical signs of hepatogenous photosensitivity, including:

- Discharges from the eyes and nose
- Sunburn and edema of skin on the muzzle, eyes and nose progressing to tissue death

Examination after death may reveal liver inflammation and lesions. Small bile ducts may be obstructed. Kidneys and adrenal glands may also show lesions.

Researchers have reported toxic signs in sheep (swellhead) after several weeks of grazing; in other cases, signs appeared after only a few days. If not removed from pastures, up to 100 percent of affected animals can die.

Horses are also susceptible to kleingrass toxicity, but unlike sheep, they do not exhibit classical signs of photosensitivity. Therefore, detecting poisoning early may be difficult. A serum chemistry profile can be very helpful.

Poor body condition and weight loss (i.e., "hard keepers") may be the only early signs. With long-term exposure, liver damage may be lethal in horses.

144

Integrated management strategies

Cattle are not reported to be susceptible to kleingrass-induced photosensitivity. Therefore, they are the best species to graze kleingrass pastures. If other livestock species have access to kleingrass, use caution. Monitor animals closely, and limit the time spent grazing kleingrass, especially when the grass is green and growing. Rotational grazing to other grass species may help.

Most animals showing clinical signs recover when kleingrass is removed from the diet. Horses should not be fed kleingrass.

Seed head ↗

Pasture ↘

↙ Whole plant

African Rue
Peganum harmala

A member of the caltrop family, African rue is a bright green, succulent, perennial herb growing from a woody base. It is bushy, many branched and about 1 foot tall when fully grown.

The leaves are alternate, hairless and divided into narrow segments. The flowers consist of five white petals and are present from April to September, with seed pods developing in May to October.

Distribution and habitat

African rue is native to the deserts of Africa and southern Asia. First recognized in the United States in 1935 on a section of land near Deming, NM, it has since spread onto dry rangelands in Arizona, southern New Mexico and western Texas. Regions: 7, 8, 10.

Toxic agent

African rue, which contains at least four poisonous alkaloids, is toxic to cattle, sheep and probably horses.

The seeds and fruit of the plant are the most toxic; a lethal dose is 0.15 percent of the animal's body weight. Young leaves are somewhat less toxic than seeds, with a lethal dose of about 1.0 percent of the animal's weight, while mature leaves are less toxic. Dry leaves are apparently nontoxic.

Livestock signs

Signs of chronic poisoning include:

- Loss of appetite
- Listlessness
- Weakness of the hind legs
- Knuckling of the fetlock joints

Acute conditions produce these signs:

- Stiffness
- Trembling
- Incoordination
- Frequent urination

The body temperature of poisoned animals is usually subnormal. They salivate excessively, wetting the lower jaw and muzzle. Postmortem examination may reveal hemorrhages on the heart or liver. Acute poisoning usually is caused by eating seeds.

Integrated management strategies

African rue is extremely unpalatable. Animals eat it only if they are starved or suffering from severe mineral deficiencies.

If poisoning occurs, it is usually in spring and summer. When possible, avoid pastures infested with African rue during these times. Remove livestock known to be grazing young leaves or seeds of African rue from the area. If given good-quality feed and water, chronic poisoning cases generally recover.

Chemical control treatments for African rue should target problem areas such as roadsides, livestock pens and traps, and areas around oil fields.

↖ **Flower**

Seedling and root ↗

Whole plant ↓

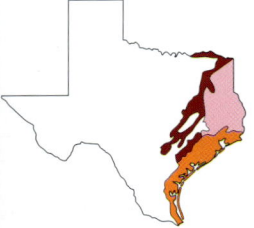

Beefsteak-plant, Perilla Mint
Perilla frutescens

Beefsteak-plant is a coarse, aromatic annual, often purple or tinged with purple. Its branching, erect, square stem is usually 30 to 45 inches tall, but may be as tall as 5 feet.

The opposite, simple leaves are oval and pointed, have a pronounced serrate or incised edge and may be up to 6 inches long.

The small white or purple flowers are arranged in terminal spikes up to 6 inches long arising from the leaf axils. Ornamental varieties often have a stronger purple tinge and a ruffled edge to the leaves.

Distribution and habitat

Beefsteak-plant was introduced from the Orient, escaped cultivation and is widespread across the eastern United States. These plants require ample water and are found in east Texas and into Arkansas, Oklahoma and Louisiana. Perilla mint can grow in dense stands and is usually found at the edge of damp woods and in open spots along streams and seepage areas (part sun/part shade). Regions: 1, 2, 3.

Toxic agent

Perilla mint is known to contain several toxic furan ketones. These compounds are toxic to the lungs of cattle, sheep, horses and laboratory animals. The toxins are present throughout the entire plant, but are concentrated in the inflorescence (flower cluster); most cases of poisoning occur after flowering, in late summer or early fall.

Hay containing the mature plant is potentially toxic, but does not pose a great hazard.

Livestock signs

Clinical signs are:
- Labored breathing
- Open-mouth breathing
- Expiratory grunt
- Death

Affected animals appear to have pneumonia and suffer from oxygen deprivation because of pulmonary edema and emphysema. Attempts to move or treat them should be slow and easy.

Integrated management strategies

Atropine can aid recovery, but the stress of treatment may make it counterproductive. Do not graze pastures with a large amount of perilla mint in the late summer or fall when other forage is short.

↖ **Whole plant**

Leaf ↗

Whole plant ↓

Abnormal Leaf-flower
Phyllanthus abnormis

Abnormal leaf-flower is a small, bushy annual or short-lived perennial herb. The plant is a member of the spurge family and is also known as sand leaf-flower.

It branches primarily from the base and grows to 6 to 12 inches tall. It flowers from May to August.

Distribution and habitat

Abnormal leaf-flower usually grows in deep sands or sandy soils in the Rolling and High Plains, Trans-Pecos and South and South-central regions in Texas. It also occurs in Oklahoma and Mexico. Regions: 2, 3, 4, 5, 6, 7, 8, 10.

Toxic agent

Abnormal leaf-flower is toxic to sheep, cattle and goats. Sheep and goats are more resistant than cattle. The toxic agent involved is unknown.

Feeding experiments have shown that the amount of plant needed for it to be toxic varies considerably depending on where the plants grow, how much toxicity is lost when they are dried, and differences in susceptibility of the various test animals.

In goats, the least amount of fresh plant material necessary to produce death was 1.5 percent of the animal's weight.

Livestock signs

Poisoned calves showed:

- Listlessness for several days
- Anorexia
- Ceaseless walking
- Periods of nonbelligerent charging about

Diarrhea and occasional rectal prolapse were observed in some animals. Exhaustion was followed by prostration and death.

Postmortem examination may show an orange or yellow liver and small purplish hemorrhagic spots on the heart or the mesentery (the membranes enfolding some internal organs).

The plant is also suspected of causing hepatogenous photosensitization.

Integrated management strategies

As with most spurges, abnormal leaf-flower is unpalatable to livestock. As such, grazing management practices ensuring enough desirable range forage significantly reduce livestock losses to abnormal leaf-flower.

Phosphorus supplementation may also reduce the incidence of poisoning from this plant. If signs are noticed, move livestock to a fresh pasture to alleviate the problem.

↑ Seedling

Seed pod ↗

Whole plant ↓

Pokeberry, Pokeweed
Phytolacca americana

Pokeberry is a smooth, shiny plant arising from a large perennial rootstock. Its stalks become purple-red with age and are usually 5 or 6 feet tall but may be as tall as 10 feet.

The large leaves, up to 4 inches wide and 10 inches long, are entire (the edges have no notches or indentations) and alternate. The small white or pinkish flowers are arrayed in drooping spikes and give rise to juicy, purple-black berries.

Distribution and habitat

Pokeweed is found in the eastern two-thirds of Texas and is reported from Maine and Ontario to Florida and California. It usually grows in disturbed, sandy soils and is often found in bulldozed brush piles. Regions: 1, 2, 3, 4, 5, 6, 7, 8.

Toxic agent

Saponins are concentrated in the rootstock and young leaves of pokeweed. All species of animals have been affected by consumption of the plant. Pigs uprooting the plants and consuming the rootstocks are the most likely to be poisoned.

Young leaves are eaten by humans as greens, but the water must be changed during cooking to remove the toxins.

Livestock signs

This plant is a gastrointestinal irritant, with clinical signs of poisoning occurring within a few hours of consumption:

- Abdominal pain
- Vomiting
- Diarrhea (sometimes with blood)
- Death

Consumption of the leaves usually produces only a transient gastrointestinal disease in ruminants. Cattle often consume small amounts of mature leaves with no ill effects.

Integrated management strategies

Do not place pigs in a pen containing well-established plants with large rootstocks. Short pastures containing old brush piles with large amounts of pokeberry and little other forage should not be used for livestock.

↖ Seedling

Flower, fruit
and leaf ↗

Whole plant ↓

153

Knotweed
Polygonum spp.

There are 18 species and several varieties of knotweeds in Texas. These plants are herbs or herbaceous-textured vines without tendrils. They have alternate, usually lance-shaped leaves.

The branching stems are erect or ascending; the enlarged nodes often take root where they contact bare ground.

The very small, white to pinkish-red flowers usually form loose spikes.

Distribution and habitat

Knotweeds are found in all vegetational areas of Texas. They are usually found in wet areas and around ponds and streams, and are therefore more abundant in the eastern part of the state. Regions: 1, 2, 3, 4, 5, 6, 7, 8, 9, 10.

Toxic agent

The toxic agent is not known. Consumption of these plants has been associated with primary photosensitization, and until research is completed to show otherwise, all species should be considered as potentially toxic. Vegetation from species associated with field cases contains photodynamic compounds not yet identified. Feeding trials with knotweed seed in grain screenings did not cause problems or disease.

Livestock signs

The clinical signs are those of sunburn and include:

- Photophobia (animals try to stay in the shade)
- Reddening of light colored skin
- Crusting and cracking of non-pigmented skin
- Sloughing of affected skin

Young calves on cows with severe udder lesions may not be allowed to nurse and can suffer significant weight loss.

Integrated management strategies

These plants are widespread, but are seldom consumed to any extent except in severe drought or overgrazed pastures.

Feeding hay can prevent consumption and toxicity. Remove affected animals from the pasture and give them shade, feed and water.

↖ Leaf
and nodes

Flower ↗

Whole plant ↓

Mesquite
Prosopis glandulosa

Mesquite is a small to medium-height tree or shrub, thorny and single stemmed or branching near the ground. The leaves are deciduous and located alternately along the stems.

The fruits are loosely clustered pods (beans) up to 8 or 10 inches long and may be abnormally abundant in drought years.

Distribution and habitat

Generally found throughout Texas, mesquite is common on dry ranges and in washes and draws at low elevations in the Trans-Pecos region. It is found from California to Texas, Kansas and Mexico. Regions: 1, 2, 3, 4, 5, 6, 7, 8, 9, 10.

Toxic agent

Mesquite beans primarily affect cattle, although goats have also been affected. Sheep are reportedly resistant. Horses that eat the beans may be susceptible to impaction colic.

In some ways, the syndrome produced by a diet of mesquite beans is best considered a nutritional problem. Mesquite beans have a high sugar content that, together with inadequacy in other dietary factors, alters rumen microflora, inhibits cellulose digestion and contributes to rumen stasis and impaction. B-vitamin synthesis is inhibited. Ketosis and starvation follow in severe cases.

Livestock signs

The signs of the disease—jaw and tongue trouble—develop gradually, usually after cattle have been eating beans for 2 or more months. Animals may lose 50 percent of their weight.

Afflicted cattle may salivate, chew continuously, sometimes with nothing in their mouths, and hold their heads to one side as if chewing is painful. About 25 percent of affected animals have a partial paralysis of the tongue, which protrudes 1 to 4 inches from the mouth. At least 10 percent of poisoned animals have swelling under their jaws or tongue, and some have noticeably enlarged salivary glands.

Signs may include:
- Loss of appetite
- Rapid weight loss
- Nervousness
- A wild expression
- Bulging eyes
- Death

Horses with impaction colic will stand in a humped position and may kick at their abdomens.

156

Integrated management strategies

Most cases of poisoning occur in pastures where many mesquite beans accumulate, such as where pack rats store them. Prevent cattle from consuming beans as 60 percent of the diet for more than 60 days without adequate high-quality roughage.

If the disease has not progressed too far, three of four animals survive if given high-quality ground feed and rumen inoculation with fresh rumen fluid from an unaffected animal. Stocking cattle and sheep together can reduce cattle losses, because the beans apparently do not affect sheep.

There are many specific chemical control recommendations for mesquite. For herbicides and rates, see Extension publication B-1466, *Chemical Weed and Brush Control Suggestions for Rangeland.*

Beans ↗

Whole plant ↓

Wild Plum, Cherry, Peach
Prunus spp.

Fourteen species and several varieties of *Prunus* are listed in Texas. These range from shrubs less than 6 feet tall to trees about 100 feet tall. Most are usually not over 15 feet.

They are usually deciduous, and winter buds have many overlapping scales. The leaves are seldom entire and are usually toothed along the margins.

The flowers have five white, pink or red petals and are solitary or arranged in tight bundles or spikes. The fruit is single-seeded and fleshy.

Distribution and habitat

Prunus species are widespread, and two or more species grow in each vegetational area of Texas. Some grow in open areas, while others are found as undergrowth in wooded sites. Many of the shrubby species grow in mottes in fields and pastures or along fence rows. Regions: 1, 2, 3, 4, 5, 6, 7, 8, 9, 10.

Toxic agent

Cyanogenic glycosides contained in the seeds and leaves of *Prunus* plants are broken down to free cyanide in the gastrointestinal tract or in damaged plant material.

All livestock species are susceptible to cyanide poisoning, but most cases occur in ruminants because conditions in the rumen favor hydrolysis more than those in the acidic stomachs of other species.

Wilted or frost-damaged leaves are very hazardous, as they contain free cyanide. Poisoning is often encountered when mechanically damaged trees or shrubs are available for livestock consumption.

Livestock signs

Cyanide is one of the most rapidly acting of all poisons. Clinical signs can occur within 5 minutes after consumption of the plant. Death may occur within 15 minutes or several hours. The signs of poisoning may include:

- Salivation
- Labored breathing
- Muscle tremors
- Incoordination
- Bloating
- Sustained contraction of voluntary muscles
- Bright red venous blood
- Convulsions
- Death from respiratory failure

Integrated management strategies

Do not allow animals access to damaged leaves until the leaves are completely dry. Remove cattle from pastures when *Prunus* is being bulldozed to prevent access to the damaged plant material.

Rapid intravenous treatment with sodium nitrite and then sodium thiosulfate is effective in animals showing clinical signs.

↖ Leaf

Flower ↗

Whole plant ↓

159

Paperflowers
Psilostrophe tagetina, P. gnaphaloides

There are two species of paper-flower in Texas. Woolly or hairy paperflower *(P. tagetina)* is similar to cudweed paperflower *(P. gnaphaloides)* in foliage, manner of growth and general appearance, but usually is much woollier. Both species affect livestock the same way.

Paperflowers are erect, perennial herbs of the sunflower family. Stems and leaves are thinly hairy. Leaves are alternate and narrow, and generally decrease in size as they move upward on the plant. Lower leaves form a rosette at the plant base.

Plants flower continuously from spring through first frost. Flower heads are yellow, occurring in loose to dense clusters at branch ends. The flower petals become pale and papery with age, hence the name paperflower.

Distribution and habitat

Paperflowers are typically found on open, dry ranges in semiarid regions of Texas, Utah, New Mexico, Arizona and Mexico. Both species are very common in western Texas. Regions: 6, 7, 9, 10.

Toxic agent

The toxic agent is a sesquiterpene lactone. The plant is more toxic in the young, green stage than after maturity. In most cases, paperflowers are considered poisonous only to sheep, although cattle may be susceptible.

Livestock signs

Paperflowers are relatively palatable to sheep, which can graze the plant for up to 2 weeks before poisoning signs appear. Signs from paperflower toxicity include:

- General discomfort and uneasiness
- Weakness of the hind legs
- Regurgitation of food or greenish liquid

At first, sheep may appear in good condition, but show some incoordination and stumbling when running. Later they become sluggish, lose their appetite and cough violently, causing them to regurgitate a greenish liquid. Eventually they become depressed and emaciated and die after a week or more of partial coma.

Integrated management strategies

No livestock treatment is specified for paperflower poisoning, although most animals recover completely if moved to a noninfested pasture or put on feed as soon as signs appear.

To minimize paperflower consumption, supplement the diet

with a high-quality protein. Because paperflower can be grazed for about 2 weeks before signs occur, manage grazing to reduce plant density by placing many animals on pastures for short periods. Rotate grazing between infested and noninfested pastures to help control losses. Another successful strategy is to graze the pasture first with cattle only, then allow access by sheep after the paperflower has matured.

← Seedling

Flower ↗

↙ Whole plant ↘

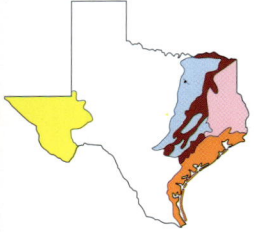

Bracken Fern
Pteridium aquilinum

Bracken fern is a coarse plant with slender, woody, branching, underground rhizomes that allow it to form large colonies. Its triangular, single fronds are erect-ascending and three times compound.

Like other ferns, bracken reproduces by spores produced on the back of the fronds.

Distribution and habitat

Bracken fern is found in the eastern third of Texas and the high mountains (above 7,000 feet) of Jeff Davis County. Many varieties extend across much of the United States, wherever there is adequate moisture. The plant colonies are usually found on hillsides at the edge of woodlands and thickets in partial shade. Regions: 1, 2, 3, 4, 10.

Toxic agent

These plants have two types of toxic agents. The first is a thiaminase, which is responsible for central nervous system disease in horses. The diet must contain 20 to 25 percent bracken fern for 3 weeks or more for horses to demonstrate clinical signs. The thiaminase is detoxified by rumen microbes, so ruminants are not affected by the plant in this way.

Bracken fern's other toxin is a norsesquiterpene lactone called ptaquiloside, and it causes disease in ruminants after several weeks or months of intake at 20 to 25 percent of the diet.

Livestock signs

In horses, clinical signs of thiamine deficiency include:

• Anorexia

• Weight loss

• Incoordination

• Muscle tremors

The disease in cattle is marked by severe bone marrow suppression and bladder tumors, which result in a hemorrhagic syndrome. Clinical signs include:

• Hemorrhages of the mucus membranes

• Weight loss

• Bloody feces

• Bloody urine

Integrated management strategies

Bracken fern poisoning is not a big problem in Texas, as there is a limited amount of the plant and a large dose is required for poisoning.

Livestock should not be forced to consume the young plants in the spring when they are palatable. Horses with clinical signs respond to thiamine injections and removal of bracken fern from the diet. Prevention is the only treatment for ruminants.

162

Leaf ↑

Whole plant ↓

Oak
Quercus spp.

There are 39 species and many varieties of oak in Texas. They range from shrubs about 3 feet tall to very large trees.

The leaves, usually deciduous and stemmed, are alternate, and with or without toothed margins or deep lobes.

The fruit is one seeded, enclosed in a shell forming a nut or acorn, and is seated in a cup that envelopes the whole nut or covers only its base.

Distribution and habitat

Several species of oak are found in each of the vegetational areas of Texas and about 500 species are across the Northern Hemisphere. Their habitat ranges from dry rocky slopes to wet bottomland to heavy sands. Most species are specific to a certain habitat type. Regions: 1, 2, 3, 4, 5, 6, 7, 8, 9, 10.

Toxic agent

The toxins in oak are complex compounds called gallotannins. These compounds are toxic to cattle, sheep, goats, horses and dogs. The free toxin is present in hazardous concentrations in the buds, flowers, young stems and acorns, but mature leaves are not toxic.

Most poisoning by shrubby species (shinnery) occurs in the spring when livestock consume many young leaves, buds, stems and/or flowers. Poisoning from large trees is most often the result of livestock consuming a large amount of acorns in the fall and early winter. However, there are cases each year in which cattle consume the young leaves from tall trees blown down during a wind or hailstorm.

As little as 6 percent of an animal's body weight of dry plant material may be enough to cause oak poisoning.

All species of oak should be considered as potentially toxic, but live oak and white oak are seldom involved.

Livestock signs

Signs of poisoning usually are seen a week or more after animals consume a lethal dose and include:

- Depression
- Constipation with blood
- Rough hair coat
- Abdominal pain
- Frequent urination, then no urination

Some cows will have subcutaneous edema between the rear legs above the udder. Many have a bloody froth from the nose after they have been handled.

Horses and dogs may die from liver failure after a brief period of depression.

Integrated management strategies

Oak poisoning can be prevented by offering supplemental ruminant feed (4 pounds per head per day) containing 10 percent hydrated lime, 6 percent fat, 30 percent alfalfa and 54 percent cottonseed meal starting before the buds are set in the spring. The same feed has helped prevent acorn poisoning, but does not work as well because the acorns are available for a longer period.

Prevention is best accomplished by moving cattle out of pastures severely infested with shinnery for about a month in the spring when buds are present. Sheep and goats are only poisoned when oak makes up the majority of their diet. Make other forage or browse available, especially in drought, to prevent poisoning.

Shinnery oak can be controlled with a broadcast treatment of Spike 20P® applied at 0.75 to 2.0 lb. a.i/acre, with the actual rate depending upon species. Individual plants may be controlled with Velpar L® applied at a rate of 4 ml per inch of stem diameter.

Acorn ↗

Whole plant → ↘

↙ Bud

165

Buttercup
Ranunculus spp.

There are 18 species of *Ranunculus* in Texas. These are perennial or annual herbs with a sharp, bitter taste.

The stem leaves are alternate, with palmlike veins, and are deeply lobed or dissected. The basal leaves usually have a distinctly different shape.

Flowers are arranged in fan-shaped clusters. They usually have five glossy yellow petals and give rise to a small, dry fruit.

Distribution and habitat

There are two or more species of buttercup in every vegetational region of Texas. However, significant populations are usually found only in the eastern third of the state.

Virtually all of the species require ample water and are found in seeps, mud flats, along ditches or in standing, shallow water. Regions: 1, 2, 3, 4, 5, 6, 7, 8, 9, 10.

Toxic agent

All species are thought to contain a glycoside at various concentrations that is converted to protoanemonin, which acts as a blistering agent. The levels of glycoside increases greatly as the plants mature and reach the flowering stage.

Because protoanemonin is not stable, the plant is not a problem in hay. Although the toxin content varies widely within and among species of buttercup, a large amount of plant material is usually required to cause clinical signs with the species growing in Texas.

Livestock signs

The signs of poisoning are those of severe gastrointestinal irritation and include:

- Red and/or ulcerated oral tissues
- Salivation
- Blood-tinged milk
- Diarrhea
- Abdominal pain
- Depression or excitation
- Convulsions
- Death

Most cases of buttercup poisoning in Texas are not life threatening. Horses consuming buttercup can die from colic.

Integrated management strategies

Poisoning can usually be prevented by not forcing animals to consume buttercup at flowering. Some pastures must be vacated to prevent diarrhea. These may be used for hay if enough forage such as ryegrass is mixed with the buttercup.

↖ Leaf

Flower ↗

Whole plant ↘

Castor-bean
Ricinus communis

Castor-bean is a coarse, hairless, annual herb growing from 3 to 15 feet tall with a single stem below and numerous ascending branches above. The large, alternate, stemmed leaves are up to 20 inches long and have seven to nine deeply toothed, palmlike lobes.

Some ornamental varieties are showy with purplish stems and leaves. The flowers lack petals but the centers are reddish. Mottled seeds are encased in three-celled capsules in terminal spikes.

Distribution and habitat

Castor-bean has been cultivated as an oilseed crop and grown as an ornamental in many areas of the state. It has become established as an escapee in some areas and can be found in fields, gardens and along some rivers in central and west Texas. Regions: 1, 2, 3, 4, 5, 6, 7.

Toxic agent

The seeds contain ricin, one of the most toxic compounds known. Fortunately, this compound is not readily absorbed through the wall of the intestinal tract. It is a protein and can be denatured (its properties can be changed) by heat.

Whole or poorly masticated seeds are not very toxic, but toxicity increases as the seed is finely ground. Horses are poisoned by about 0.01 percent of their weight in seeds; cattle, sheep and pigs must consume about 0.25 percent of their body weight. Animals have been poisoned by grain contaminated with castor-bean seeds.

Livestock signs

There is often a delay of several hours to days between consumption of castor-bean seeds and onset of clinical signs, which are related to severe gastric inflammation and upset. Signs can include:

- Anorexia
- Depression
- Depressed rumen function
- Abdominal pain
- Colic in horses
- Diarrhea
- Weakness

Integrated management strategies

Castor-bean poisoning is rare in North America. Animals seldom consume seeds from mature plants, even though they are available and other forage is scarce. Remove seed clusters from ornamental plants before they are mature to prevent consumption and poisoning by pets.

Feed containing castor-bean seeds should not be fed to livestock.

↑ Leaf

Flower ↗

Whole plant ↘

↙ Seed pod

Black Locust
Robinia pseudo-acacia

Black locust is an introduced tree up to 50 feet tall that often forms colonies by root sprouts. Spines on the young stems are unbranched and resemble large rose thorns. Seven to 19 oval leaflets are arranged opposite with the entire leaf up to 6 inches long.

The fragrant, abundant flowers have white petals and a banner with a yellowish center and are arranged in hanging clusters 4 to 8 inches long. The pods are flat, thin and brown and may persist through the winter.

Distribution and habitat

Black locust is extensively planted in shelterbelts and as an ornamental. It can be found in all areas of Texas except the Rolling Plains and the Trans-Pecos. It is probably native to the southeastern United States. Regions: 1, 2, 3, 4, 5, 6, 7, 9.

Toxic agent

The toxic agent of black locust is robin, a protein toxin. All parts of the plant except the flower are toxic.

Horses, cattle, sheep and poultry have been poisoned. Horses are most often poisoned and are the most susceptible species. Horses that consumed as little as 0.04 percent of their weight in bark showed signs of poisoning in 1.5 hours.

Livestock signs

Signs of poisoning are similar in all species and may include:
- Anorexia
- Depression
- Diarrhea
- Weakness (posterior paralysis in cattle and horses)
- Cold extremities
- Weak pulse
- Irregular heartbeat

In fatal cases, death usually occurs within 1 or 2 days.

Integrated management strategies

Most clinically affected animals recover after removal from the source. However, complete recovery may take several weeks, and horses often founder or develop laminitis. Do not place horses in a paddock with a black locust tree. Boredom may cause them to consume the leaves or strip the bark from the trees.

↖ Leaf

Flower ↗

Whole plant ↓

Dock
Rumex spp.

There are 15 species of *Rumex* listed in Texas. Most of these are perennial herbs with a basal rosette of leaves. The stalk is upright and has alternate, mostly entire (edges have no notches or indentations) leaves.

The small, greenish flowers are arranged in dense clusters on elongated stems. The mature flowering stalk and the three-sided fruit are usually brown at maturity. Each fruiting body contains a single black or brown seed.

Distribution and habit

At least one species of dock can be found in each vegetational area of Texas. It is among the most common weed across the United States and southern Canada. Dock grows most often on disturbed soil, and most Texas species are found in seasonally wet areas. Regions: 1, 2, 3, 4, 5, 6, 7, 8, 9, 10.

Toxic agent

Dock contains soluble oxalates, and the concentration tends to increase as the plant matures. Poisoning is uncommon, but occurs most frequently in ruminants. At a low level of intake, the rumen microorganisms adapt over a period of several days and are able to use oxalates to produce energy.

Toxicity can occur when non-adapted animals consume a large amount of oxalate. In oxalate poisoning, calcium oxalate crystals are deposited in the kidneys and in blood vessel walls.

Livestock signs

Clinical signs of poisoned animals may include:

- Anorexia
- Incoordination
- Depression
- Prostration
- Convulsions
- Death

Severely poisoned animals that live for several days can die from kidney failure.

Integrated management strategies

Good pasture or range management can prevent the sudden intake of the large amount of dock necessary to produce poisoning. Most cases of poisoning occur when these unpalatable plants are treated with a herbicide such as 2,4-D without removing the animals from the pasture. This treatment makes the plant more palatable and may also increase its oxalate content.

↖ **Flower**

Whole plant ↗ ↓

Russian Thistle, Tumbleweed
Salsola iberica

Tumbleweed is a many-branched, annual herb growing to 2 to 6 feet tall. At maturity, the plant is stiff, prickly, round and bushy. The spine-tipped leaves are oval. The stems have distinctive dark purplish striations (parallel to the stem) when the plant is young and growing.

Tumbleweed is a member of the goosefoot family.

Distribution and habitat

Tumbleweed is found in every region of Texas except the Piney Woods and Post Oak Savannah. It is most abundant along roads, in irrigated fields and in disturbed areas. Regions: 2, 4, 5, 6, 7, 8, 9, 10.

Toxic agent

Nitrate is the toxic agent. All ruminants are susceptible to nitrate poisoning, with cattle poisoned most often. Plants with more than 1.0 percent nitrate are dangerous; animals may die if they have consumed as little as 0.075 percent of their weight in nitrate.

Environmental factors often influence nitrate. For example, nitrate poisoning is more likely if the plant grows in soils high in nitrogen, such as in livestock pens or fertilized areas. Excessive shade, lack of water, and stress or physical damage may also increase nitrate levels.

Livestock signs

Animals with acute nitrate poisoning are often found dead with no previous history of illness. Less acute nitrate poisoning signs occur in this order:

- Weakness
- Unsteady gait
- Collapse
- Shallow and rapid breathing
- Rapid pulse
- Dilated pupils
- Recovery
- Delayed abortion, or
- Coma
- Sudden death

Unpigmented parts of the body such as the whites of the eye, the tongue and lips may have a blue-brown discoloration; the blood may be a chocolate brown color.

Integrated management strategies

Avoid areas infested with this plant during drought, after a period of extended cool, cloudy weather, or after a heavy application of nitrogen fertilizer. Rations high in carbohydrates also help reduce losses from nitrate poisoning.

Keep poisoned animals quiet, and administer methylene blue intravenously. Generally, use a 1 to 4 percent solution containing 5 percent dextrose at a rate of 1

gram of methylene blue per 250 pounds of animal weight. Be certain of the diagnosis of nitrate poisoning before treating with methylene blue.

↖ Seedling

Stem ↗

Whole plant ↓

Lanceleaf salvia
Salvia reflexa

Lanceleaf salvia, also called Rocky Mountain sage, grows to about 2 feet tall, attaining a multi-stemmed, bushy stature. The square stems branch opposite each other at an upward angle.

The narrow, lance-shaped leaves up to 2 inches long are also arranged opposite each other on short stems. Asymmetrical, blue flowers grow on spikes in whorls of two or three.

Distribution and habitat

These plants are found in the western two-thirds of Texas. They grow in dry fields, gravel-clay flats, slopes and rocky soils where there is little competition. Regions: 4, 5, 7, 8, 9, 10.

Toxic agent

The toxic agent is unknown. Reported cases of poisonings in the United States are limited to cattle and horses that consumed contaminated hay. Experimental feeding trials have shown that sheep are also susceptible.

Although the plant has been proven toxic, it is not known how much of the plant material must be eaten to cause toxicity. In one confirmed case, alfalfa hay contained about 10 percent lanceleaf salvia.

Livestock signs

Salvia poisoning is not common, and only general clinical signs are reported:

• Muscular weakness
• Diarrhea
• Colic

Animals that die after consuming contaminated hay show post-mortem evidence of gastrointestinal inflammation and liver necrosis.

Integrated management strategies

There are no documented reports of this plant causing poisoning under range or pasture conditions, although it has been suspected. Nevertheless, it has been proven toxic when consumed as a contaminant in hay and should therefore be considered potentially toxic to grazing animals. Look for this plant in hay as well as in hay fields before mowing.

↖ Seed pods

Flower ↗

Whole plant ↓

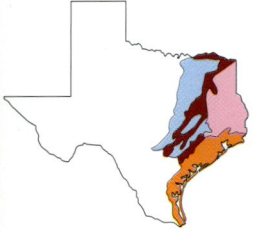

Chinese Tallow Tree
Sapium sebiferum

Chinese tallow is a fast-growing weedy tree with milky sap. It grows up to 30 feet tall and often spreads by root sprouts. Its slender limbs and branches droop and are easily broken.

The leaves are hairless, alternate with smooth margins and have diamond-shaped blades that are shorter than the petioles. They turn bright yellow, orange or red in the fall. The flowers have no petals and grow in 2- to 6-inch drooping spikes at the end of each branch.

The walls of the three-celled fruit fall readily at maturity, leaving three chalky white seeds, which may remain attached through the winter. These nutlike seeds have a hard coat covered by tallow that becomes black with weathering.

Distribution and habitat

Chinese tallow was introduced from Asia and is planted widely as an ornamental. Birds disperse the seeds, and it has escaped in the southeastern part of Texas, where it can be a significant invading woody species. Regions: 1, 2, 3, 4.

Toxic agent

The toxic agent is unknown. The terminal leaves and green fruit have a strong purgative effect on the bowels of cattle.

The consumption of 1 percent of an animal's weight in green plant material can produce clinical signs within 12 to 14 hours. Losses may occur when cattle are forced to consume significant amounts of the plant in the seedpod stage. Sheep and goats are not affected significantly.

These trees are more significant as a cause of decreased forage production than as a toxic plant.

Livestock signs

The signs of poisoning are associated with gastrointestinal disorder and may include:

- Diarrhea (sometimes with free blood)
- Anorexia
- Listlessness
- Weakness
- Dehydration

Integrated management strategies

Animals consuming the plant and suffering from diarrhea often recover when it is removed from the diet. Chinese tallow is unpalatable and is not consumed when good grazing management practices are followed. Pastures with many tallow trees should not be severely overgrazed, especially in early summer when the trees are putting on fruit.

178

Severe populations of Chinese tallow may be controlled with broadcast applications of Grazon P+D® (2.5 pounds a.i./acre) or Tordon 22K® (0.5 pound a.i./acre) in the spring or fall.

Individual plants may be controlled with spot applications of Velpar L® at 4 ml per 1 inch of stem diameter or with a basal stem treatment of 25 percent Remedy® plus 75 percent diesel fuel oil.

↖ Flower

Leaf and fruit ↗

Whole plant ↓

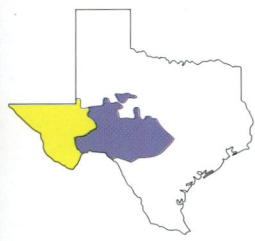

Threadleaf Sartwellia
Sartwellia flaveriae

Threadleaf sartwellia is a bushy, stiff-branched annual of the sunflower family that typically grows from 6 inches to 2 feet tall. It is the only species in this genus recorded in Texas.

Its numerous leaves are threadlike, sticky and 1 to 2 inches long. The yellow composite flowers emerge from August through October.

This plant is easily confused with perennial broomweed.

Distribution and habitat

Threadleaf sartwellia grows in alkaline or gypsiferous soils and can form dense stands. Found in the Trans-Pecos and western Edwards Plateau regions of Texas, it is especially common along highways in Reeves, Loving and Ward counties. Regions: 7, 10.

Toxic agent

The toxic agent is unknown. Poisoning has been demonstrated only in goats. The toxic dose is 1.5 to 3.0 percent of the animal's weight in green plant material.

Livestock signs

Signs documented in rangeland situations include:

- Gradual loss of weight despite normal appetite
- Distended abdomen

In feeding experiments, goats ate the plant readily until signs of anorexia appeared within 6 to 28 days. They died after continued forced feeding.

Postmortem examination has revealed kidney damage and liver necrosis with numerous grayish yellow areas throughout.

Integrated management strategies

Remove goats with poisoning signs from infested areas and give them good-quality feed and water.

Good grazing management practices that provide a variety of forage plants should limit threadleaf sartwellia consumption.

Flower ↑
Whole plant ↓

181

Texas Groundsel,
Texas Squaw-Weed
Senecio ampullaceus

Texas groundsel is a cool-season annual herb. It grows to 12 to 30 inches tall. The plants are often whitish with hair, but can be nearly hairless.

The unlobed, clasping leaves gradually reduce in size toward the top of the plant. Showy yellow flowers are produced in the spring.

The seedling, or winter rosette, often has a purplish cast to the underside of the leaves, especially on the midrib.

Distribution and habitat

Texas groundsel is found in the eastern half of the state. It is abundant in sandy soils and may be a predominant species in freshly cleared forest. Regions: 1, 2, 3, 4, 5, 6, 7.

Toxic agent

The toxic agent of these plants has not been established, nor have experimental feeding trials proven its toxicity.

In field cases, cattle consuming Texas groundsel developed clinical signs and lesions typical of pyrrolizidine alkaloid poisoning, as occurs with other *Senecio* species. Clinical signs appeared several months after it was ingested; the plants were long dead when the animals became ill.

The dead animals had classical liver cirrhosis identical to that produced by pyrrolizidine alkaloids. Tests on many normal-appearing herd mates of the dead animals showed that they also had liver damage.

Llamas pastured with the plant for months at a time have also succumbed to terminal liver cirrhosis.

Livestock signs

The clinical signs of poisoned cattle and llamas may include:
- Anorexia
- Depression
- Weight loss
- Aggression
- Death

These animals die because of liver failure.

Integrated management strategies

Cattle are more likely to consume the young plants while they are still in the rosette form in late fall and winter. Poisonings during this period can be reduced by providing adequate hay and supplemental feed.

Because llamas seem to be either more sensitive to the plant or less selective grazers than cattle, eliminate the plants from their pastures. Sheep tolerate a greater intake of pyrrolizidine alkaloids than other animal species and may be used to remove the young plants.

Heavily infested pastures may be treated with broadleaf herbicides such as 2,4-D or Grazon P + D®.

Flower ↗

Leaf ↘

↙ Whole plant

Threadleaf Groundsel, Senecio
Senecio douglasii

Threadleaf groundsel is a many-stemmed evergreen composite. The stems and leaves are gray-green. The leaves are divided into three to seven segments and may be hairy or nearly smooth. The stems are herbaceous, although somewhat woody at the base, and may have variable hairiness.

Showy yellow flowers emerge from March through November.

Distribution and habitat

Threadleaf groundsel is a common range plant in Colorado and Utah, and south to Texas and Mexico. It is common in the grassland areas of western Texas. Disturbance and overgrazing cause it to increase in abundance. Regions: 6, 7, 8, 9, 10.

Toxic agent

Threadleaf groundsel owes its toxicity to pyrrolizidine alkaloids. Stress from lack of water causes the plant to increase in alkaloid content.

Cattle and horses are about equally sensitive to this plant. Sheep and goats are more resistant, generally requiring up to 10 times the amount for the same effect as in cattle and horses.

Generally for acute poisoning, cattle and horses must eat a dose of 1 to 5 percent of their weight in threadleaf groundsel over a few days. This type of poisoning is rare under range conditions.

Most losses are from chronic poisoning, which occurs when cattle and horses consume as little as 0.25 percent of their body weight.

Livestock signs

Often up to 6 months elapse between consumption of this plant and the appearance of chronic signs. During this period, animals may even gain weight and appear thrifty.

The first signs of poisoning include:

- Standing apart from other animals
- Depression and sluggishness
- Lacking appetite
- Weight loss

Signs of the advanced stage:

- Continuous walking, sometimes without avoiding objects
- Sudden nervous appearance upon disturbance
- Frequent voiding of small amounts of urine
- Bile-stained (yellow) feces
- Rectal prolapse (cattle)
- Skin swollen with excessive fluid and possibly emitting a sweetish, unpleasant odor
- Conversely, death occurring quickly or quietly after a period of depression

In the advanced stage, the animals may remain relatively quiet or become agitated and dangerously aggressive.

Examination after death may reveal a hardened liver (possibly with mottled coloration). The gall bladder may be distended, frequently to an enormous size.

Integrated management strategies

There is no treatment for pyrrolizidine alkaloid poisoning because the liver damage is severe, progressive and permanent.

Management practices that improve range condition will reduce losses to threadleaf groundsel. Proper mineral supplementation, especially with phosphorus, also helps.

Treat individual plants by applying Grazon P + D® (2 percent solution in water) directly to the leaves. For widespread populations, aerial or ground broadcast applications of 0.94 pound a.i./acre of Grazon P + D® or 0.25 ounce a.i./acre of Escort® have given good results.

Flower ↗

Leaf ↘

↙ Whole plant

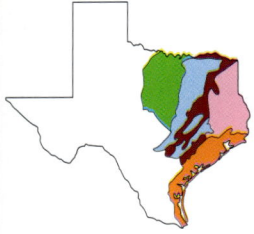

Butterweed
Senecio glabellus

Butterweed is a cool-season annual that usually grows to 18 to 20 inches tall. On rare occasions it grows to 40 inches. The basal leaves, up to 4 inches long, are deeply lobed, with oblong, lateral lobes having wavy margins.

Although the plant usually has one stalk, it occasionally has a few. It can be unbranched or branched above, depending upon moisture and soil fertility.

Numerous yellow flowers stand on short stalks, forming a large terminal cluster.

Distribution and habitat
Butterweed may be found in many of the areas of Texas, but large populations are usually present only in the eastern third of the state. It is often found in clay or heavy loam soils in disturbed areas, stream bottoms, ditches and flood plains. Regions: 1, 2, 3, 4, 5.

Toxic agent
These plants contain pyrrolizidine alkaloids that cause progressive liver cirrhosis. Consumption of large amounts of the plant can result in acute liver necrosis. However, in most instances, several months elapse between consumption and the appearance of clinical signs.

Horses, cattle and llamas have died of liver cirrhosis after being on pastures containing butterweed.

Livestock signs
Sudden death has been seen in some cases in which horses were placed on almost pure butterweed paddocks with no hay or feed.

Most animals are clinically normal for several months after consuming the plant and then demonstrate signs of poisoning, which are all related to loss of liver function and can include:

- Anorexia
- Depression
- Weight loss
- Aggression
- Death

Integrated management strategies
Animals should not be forced to consume these unpalatable plants. Poisoning has not been observed in species other than llamas if there is adequate pasture or good quality hay and feed available. Llamas may be more sensitive to the pyrrolizidine alkaloids or less discriminating in their eating habits.

Sheep are more resistant to pyrrolizidine alkaloid poisoning and may be used to "clean" the pastures while the plants are young. Severe infestations may be controlled with broadleaf herbicides such as 2,4-D or Grazon P + D® at 0.5 to 1.0 pound a.i./acre in the spring with good growing conditions.

↖ Leaf

Whole plant ↗

↓ Flower

187

Riddell Groundsel, Broom Groundsel

Senecio riddellii

Riddell groundsel is a herbaceous perennial of the sunflower family. The leaves are hairless and bright green.

Yellow flowers emerge in the fall (September to November). The plant dies back to the ground after frost.

Distribution and habitat

Riddell groundsel is a common range plant in Colorado and Utah, and south to Texas and Mexico. It is a common plant of grassland areas of western Texas. It increases in abundance with disturbance and overgrazing. Regions: 2, 6, 7, 8, 9, 10.

Toxic agent

Riddell groundsel owes its toxicity to pyrrolizidine alkaloids. When the plant is stressed from lack of water, its alkaloid content increases.

Cattle and horses are about equally sensitive to this plant. Sheep and goats are more resistant, generally requiring up to 10 times the amount for the same effect as in cattle and horses.

For acute poisoning, cattle and horses generally must eat 1 to 5 percent of their weight in riddell grounsel over a few days. This type of poisoning is rare under range conditions. Most cattle and horse losses are from chronic poisoning,

which can be caused by their consuming as little as 0.25 percent of their body weight in the plant.

Livestock signs

Often up to 6 months elapse between consumption of this plant and the appearance of chronic signs. During this period, the animal may even gain weight and appear thrifty. First signs of poisoning include:

- Standing apart from other animals
- Depression or sluggishness
- Lack of appetite
- Weight loss

The advanced stage of the disease is characterized by:

- Continuous walking, sometimes without avoiding objects
- Sudden nervous appearance upon disturbance
- Frequent voiding of small amounts of urine
- Bile-stained (yellow) feces
- Rectal prolapse (cattle)
- Skin swollen with excessive fluid and possibly emitting a sweetish, unpleasant odor
- Conversely, death occurring quickly or quietly after a period of depression

In the advanced stage, animals may remain relatively quiet or become agitated and dangerously aggressive.

Examination after death may reveal a hardened liver (possibly mottled coloration). The gall bladder may be distended, frequently to an enormous size.

Integrated management strategies

There is no treatment for pyrrolizidine alkaloid toxicity because the liver damage is severe, progressive and permanent.

Range management practices that improve range condition reduce losses to riddell groundsel. Proper mineral supplementation, especially with phosphorus, also helps.

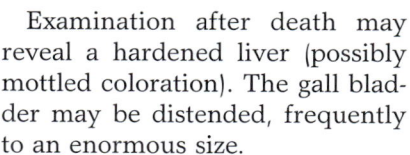

Leaf ↗

Flower ↗

Whole plant ↓

189

Lindheimer Senna
Senna lindheimeriana

Lindheimer senna is an erect, perennial, foul-smelling legume with one to several velvety stems arising from a deep, woody root.

Its compound leaves are hairy, spirally arranged and have five to eight pairs of oblong leaflets. The flowers, which have five yellow petals, are borne in upper leaf axils and are in spikes about as long as the leaves.

Distribution and habitat

This senna is found primarily in the Edwards Plateau, west through the Trans-Pecos and to Arizona and south into Mexico. It is common in shallow limestone soil, on hillsides and rocky ravines. Regions: 2, 4, 5, 6, 7, 10.

Toxic agent

The compounds responsible for the toxicity of *Senna* are unknown. Lindheimer senna is very unpalatable and is consumed only in unusual circumstances.

The plant has not been proven toxic by experimental feeding trials, but has been implicated in bovine deaths. In one case, cows from the Midwest were transported to the Hill Country and turned out in a small pasture containing a large amount of lindheimer senna. The hungry animals consumed a large amount of the plant, then displayed a clinical syndrome typical of coffee senna poisoning. Examined after death, they also had similar muscle lesions.

Livestock signs

The clinical signs are those of animals with severe muscle damage and gastrointestinal disturbance and include:

• Diarrhea
• Weakness
• "Alert downers"—not depressed, will eat, but unable to rise
• Dark urine
• Death

Few animals that go down will recover.

Integrated management strategies

Poisoning can be prevented by not forcing cattle to consume these plants. Always fill animals new to an area with good forage before placing them in pastures where toxic plants are growing.

↖ Pod

Flower ↗

Whole plant ↘

↙ Leaf

191

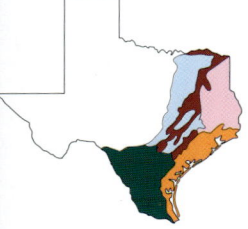

Sicklepod Senna
Senna obtusifolia

Sicklepod senna is a foul-smelling herb that grows up to 5 feet tall. It is widely spreading with numerous ascending, hairless branches.

The compound leaves are arranged spirally and usually have three pairs of symmetrically egg-shaped leaflets up to 2 inches long. One to three yellow flowers appear on short axillary stems.

The linear pods grow to 8 inches long, curve downward and contain many shiny, angular seeds.

Distribution and habitat

Sicklepod senna is found in the eastern third of Texas and extends eastward to Florida and north to Indiana and Pennsylvania. It usually grows in disturbed sandy soil and has been a troublesome weed in crops such as corn and soybeans. Regions: 1, 2, 3, 4, 6.

Toxic agent

The toxic agent of sicklepod senna is unknown. The foliage of this plant appears to be more toxic than the mature seed.

In some cases, cattle were poisoned when sicklepod foliage was included in green chop or silage, or when they were forced to consume the plant because other nutritious forage was lacking. A large amount of plant material must be consumed over several days to cause poisoning.

Livestock signs

Signs of poisoning by sicklepod senna relate to muscle damage and gastrointestinal disturbance and can include:

- Diarrhea
- Weakness
- Incoordination
- Dark, coffee-colored urine
- "Alert downers"—not depressed, will eat, but unable to rise
- Death

Most down animals remain bright-eyed and alert, and continue to eat when feed and water are placed before them. Most down animals do not recover.

Integrated management strategies

Do not cut forage for silage or green chop when it contains sicklepod senna. Take chemical control measures when these plants are growing in corn destined for green chop or silage. Cattle on short or overmature, unpalatable grass should not have access to this senna.

↖ Seedling

Flower ↗

Whole plant ↓

Coffee Senna
Senna occidentalis

Coffee senna is an erect, smooth, hairless, foul-smelling annual growing 2 to 6 feet tall.

Its ascending, branching stem has spirally arranged compound leaves with four to six pairs of leaflets. The leaflets are oval and lance shaped with pointed tips.

The linear seedpods are 4 to 6 inches long, tend to be erect and contain numerous compressed, dull brown or dull green seeds.

Distribution and habitat

Coffee senna is found in East and South Texas and extends east to Florida and north to Virginia. It usually grows in sandy or loamy disturbed soil, often in colonies around pens or shade trees rather than uniformly distributed over a pasture. Regions: 1, 2, 3, 4, 6, 7.

Toxic agent

Cattle, horses, goats and sheep have been poisoned by coffee senna. The specific chemical responsible for the toxicity is unknown, but it appears to be present throughout the plant. Generally, the unpalatable green plants are not consumed; the dried, mature seedpods are responsible for most poisonings, which usually occur after frost. Plants that have dried after being cut or pulled up have also poisoned cattle, and they are the species most often poisoned in Texas.

Livestock signs

Clinical signs of affected animals include:
- Diarrhea
- Weakness
- "Alert downers"—not depressed, will eat, but unable to rise
- Dark urine
- Death

Once an animal is down, it generally will not recover, even though it is bright-eyed and continues to eat and drink.

Integrated management strategies

The best prevention is to eliminate the plants from a pasture to be used for cattle after frost.

Mechanical removal (pulling) is quite effective in many instances, as the plants are often confined to small areas. When larger areas are involved, use of pasture may need to be deferred.

↖ Seed pods

Flower ↗

Whole plant ↓

Twinleaf Senna
Senna roemeriana

Twinleaf senna is an erect, gray, perennial herb covered with short, soft hairs. It has few to many stems arising from a thickened root.

Leaves are arranged spirally as a single pair of leaflets (hence the name twinleaf). Yellow flowers emerge from April to August with five petals about twice as long as the sepals. The flower stamens are straw-colored to light brown.

Distribution and habitat

Twinleaf senna is common in pastures and open woods on limestone soils in Central and West Texas and westward to New Mexico. Regions: 4, 5, 6, 7, 8, 9, 10.

Toxic agent

The toxin involved is unknown. Twinleaf senna is toxic to cattle, goats and horses. Sheep are more resistant, but also can be poisoned if they eat too much.

The lethal dose of dry plant material for cattle and goats is about 1 percent of their body weight eaten for 5 to 10 days. Field cases indicate that goat kids may be fatally poisoned with much less fresh plant material.

Livestock signs

Senna usually affects muscle tissue by destroying the energy-producing systems within the cells. It also causes gastrointestinal disturbances.

In goats, primarily the heart muscle is affected, and animals die suddenly of heart failure. Sometimes pale streaks are visible in the cardiac muscle.

In cattle, the large skeletal muscles are most affected.

Clinical signs include:
• Diarrhea
• Weakness
• "Alert downers"—not depressed, will eat, but unable to rise
• Dark urine
• Death

Senna-poisoned animals found already down seldom recover. Again, affected muscles may be visibly pale-streaked.

Horses and sheep poisoned with *Senna* die of liver failure. Microscopic examination of skeletal muscle reveals the same kind of damage that other species suffer, but to a lesser extent.

Integrated management strategies

The best prevention is to maintain good range condition.

Supplemental feeding programs, especially with phosphorus, also reduce twinleaf senna losses. Because of their higher resistance, including sheep in the grazing mix may reduce the amount of twinleaf senna consumed by cattle or goats.

Small populations can be controlled with Grazon P + D® or Weedmaster® herbicide (1 percent solution in water) applied directly to the leaves. Follow chemical applications with proper stocking rates and good grazing management practices.

↖ **Leaf**

Flower and beans ↗

Whole plant ↓

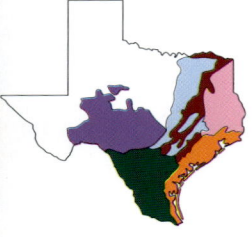

Sennabean, Drummond Sesbania

Sesbania drummondii

Sennabean is a woody-based perennial shrub of the legume family that grows to about 2 to 10 feet tall.

Twenty to 40 leaflets are arranged opposite along a main axis with leaves up to 8 inches long. Yellow flowers are borne in spikes in late summer.

The seedpod often remains closed and on the plant over the winter. It has a very characteristic four-winged structure with constrictions and cross partitions between the seeds.

Distribution and habitat

Sennabean grows in East and South Texas along the coast, extending inland along waterways. It is found in wet areas, usually in tight soil, and often grows in shallow water. Regions: 1, 2, 3, 4, 6, 7.

Toxic agent

Sennabean plants contain sesbaimide, which is concentrated in the seed. The green and flowering plants are unpalatable; only the mature, dry legumes and seed are consumed.

Animals raised in pastures with the plant learn to avoid it and are seldom poisoned. However, naive cattle, goats or sheep placed on pastures containing dried plants in late fall and winter are often poisoned.

Observations of field cases indicate that seeds are much more toxic when they first mature than 2 or 3 months later.

Livestock signs

Signs of poisoning occur within 1 or 2 days after ingestion and may include:

- Depression
- Diarrhea
- Weakness
- Rapid heart rate
- Labored breathing
- Death

Clinical signs progress rapidly; sick animals often die within 24 hours. Diagnosis is confirmed when seeds and/or seed fragments are found in the rumen contents.

Integrated management strategies

In general, good range management practices can reduce poisoning instances. Avoid placing hungry, naive ruminants in pastures containing mature *Sesbania* seedpods. Fill newly introduced animals with hay before releasing them; do not place them in heavily infested pastures without supplemental feed.

198

↑ Pod

Whole plant ↓

Bag-Pod Sesbania
Sesbania vesicaria

Bag-pod sesbania is an annual herb in the legume family that grows to 6 to 10 feet tall. The leaves consist of 20 to 40 alternate leaflets. Drooping spikes of yellowish to coral-colored flowers appear in late summer.

The beaked pods consist of two membranes, the outer one thick and the inner one papery. Each holds two or three seeds. The pods persist on the plant long after the leaves have fallen. The kidney-shaped seeds have a prominent hilum, or eye.

Distribution and habitat

These plants grow in the eastern half of Texas and extend eastward through the coastal states to North Carolina. They are usually found in well-drained sandy sites in wetter regions and in low sandy areas subject to flooding in drier regions. Regions: 1, 2, 3, 4, 5, 6, 7.

Toxic agent

Bag-pod sesbania contains sesbaimide, which is concentrated in the seed. Fresh green plants are unpalatable; only the mature dry seedpods and seeds are consumed. Animals pastured with the plant during the growing season are seldom poisoned, but naive ruminants, especially goats and cattle, are often poisoned when they are introduced to the dried plants in the fall and winter. Clinical observations indicate that newly mature seeds are more toxic than those that have weathered on the plant. The seeds of bag-pod sesbania *(S. vesicaria)* seem to be more toxic than those of sennabean *(S. drummondii)*.

Livestock signs

Signs of poisoning occur within 1 or 2 days after consumption and can include:

- Depression
- Diarrhea
- Weakness
- Rapid heart rate
- Labored breathing
- Death

Death quickly follows the onset of clinical signs, which in many cases go unobserved. Seeds and/or seed fragments are routinely found in the rumen contents of animals that die from eating this plant.

Integrated management strategies

Avoid placing hungry, naive ruminants in pastures containing plants with mature seedpods. Fill newly introduced animals with hay before releasing them, and do not place them in heavily infested pastures without supplemental feed.

200

Heavy infestations can be eliminated by mowing and prevented by good range management practices, as these plants are poor competitors.

↖ Seedling

Pods ↗

Whole plant ↓

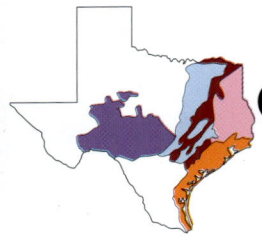

Carolina Horse Nettle, Horse Nettle

Solanum carolinense

Horse nettle is a coarse, branching perennial that grows to 1 to 3 feet high. It has spines on the stem and leaves.

The mostly oval leaves have several large teeth or shallow lobes on each side. On the underside are microscopic, star-shaped hairs.

The clustered flowers are pale-violet to white and give rise to spherical fruit about 0.5 inch in diameter. The fruit is green with light green vertical bands until maturity, when it becomes uniformly yellow.

Distribution and habitat

This plant is distributed across the eastern part of Texas and the entire eastern half of the United States. It may be found mostly in sandy soils in fields, open woodlands and waste places. Regions: 1, 2, 3, 4, 7.

Toxic agent

Horse nettle is a nightshade. Nightshades contain glycoalkaloids, which are broken down in the body to sugars and alkaloids. The toxicity may depend upon the maturity of the plants, because more toxins are present in the fruits than in the leaves.

Almost every animal species has been poisoned by nightshade, but *S. carolinense* is probably mostly responsible for cases involving cattle and horses. Hay and silage containing the mature plants have been associated with poisoning and deaths.

Livestock signs

It is thought the glycoalkaloids are responsible for the clinical signs associated with the gastrointestinal tract, and the alkaloids are associated with the signs in the central nervous system.

The signs may include:
- Anorexia
- Depression
- Excess salivation
- Diarrhea or constipation
- Trembling
- Weakness
- Colic

The star-shaped hair and seeds of this plant are readily identifiable in the gastrointestinal contents of acutely poisoned animals by microscopic techniques.

Integrated management strategies

Carolina horse nettle is not a palatable plant, and consumption by animals should not be forced. Avoid including the mature plants in hay, silage or green chop.

Chemical control strategies may be warranted in fields or pastures

202

infested with dense populations. Apply 0.6 to 0.9 pound a.i./acre of Grazon P + D® as a broadcast treatment or a 1 percent solution as an individual plant treatment when plants begin to flower in the spring.

↖ Fruit

Whole plant ↗

Flower ↓

Western Horse Nettle
Solanum dimidiatum

The leaves, petioles and branched stems of western horse nettle carry sharp spines. Most specimens of this perennial are less than 2 feet tall, but some grow to 3 feet tall.

The oval, five- to seven-lobed leaves can be up to 6 inches long. Flowers are bluish-purple to violet (rarely white) and grow in terminal clusters. They give rise to spherical fruits that are 0.75 to 1.5 inches in diameter and yellow at maturity.

Distribution and habitat

These plants are widely distributed on loamy prairies of Texas except in the far west, Panhandle and extreme eastern parts of the state. Regions: 2, 3, 4, 5, 6, 7, 8.

Toxic agent

Western horse nettle contains varying amounts of glycoalkaloids and calystegines in all parts of the plant. The highest concentrations are in the ripe fruits.

Low-level intake of calystegines over several months is probably responsible for "Crazy Cow Syndrome," a nervous condition occurring in two limited geographical areas: Real County with parts of surrounding counties, and the larger area roughly bounded by Glasscock, Menard and Taylor counties.

When cattle, sheep, goats or horses consume larger amounts, the glycoalkaloids produce signs of classical nightshade poisoning.

Livestock signs

Clinical signs of cattle with the "Crazy Cow Syndrome" include:

- Loss of equilibrium after rapid movement
- Extension of head and front legs
- Rapid eye movement
- Spontaneous but temporary recovery after rest

These signs are the result of loss of Purkinje cells of the brain. Affected animals do not return to normal. The poisoning itself is not fatal, but animals die as a result of accidents, particularly by drowning when water is nearby.

Clinical signs of classical nightshade poisoning include:

- Anorexia
- Depression
- Excessive salivation
- Diarrhea or constipation
- Trembling
- Weakness
- Colic

Integrated management strategies

Because this plant often grows in fields, take care to prevent baling fruiting plants into hay. When the plant is in round bales, cattle should not be forced to clean up the leavings, where most of the seeds will accumulate.

Do not allow the same animals to continuously graze pastures containing many western horse nettle plants, especially in the fall and winter. Rotating cattle every 30 days from this type pasture to one containing few western horse nettles has helped prevent "Crazy Cow Syndrome."

Chemical control strategies include 0.6 to 0.9 pound a.i./acre of Grazon P + D® as a broadcast treatment or a 1 percent solution as an individual plant treatment when plants begin to flower in the spring.

Leaf and flower ↗

Mature fruit →

Whole plant ↘

↙ Immature fruit

Silverleaf Nightshade

Solanum elaeagnifolium

Silverleaf nightshade is an upright, usually prickly perennial in the potato family. It normally grows 1 to 3 feet tall.

This plant reproduces by seed and creeping rootstalks. Its characteristic silver color arises from the tiny, densely matted, starlike hairs covering the entire plant. The leaves have wavy margins and are lance shaped to narrowly oblong.

The showy violet or bluish (sometimes white) flowers are followed by round yellow fruits up to 0.5 inch in diameter from May to October.

Distribution and habitat

Silverleaf nightshade is a serious weed of prairies, open woods and disturbed soils in southwestern United States and Mexico. It is occasionally found even farther north than Missouri. Regions: 1, 2, 3, 4, 5, 6, 7, 8, 9, 10.

Toxic agent

This plant has reportedly poisoned horses, sheep, goats, cattle and humans. However, sheep and goats are more resistant than cattle, and in controlled experiments, goats were not poisoned at all.

Its toxic agent is solanine. The leaves and fruit are toxic at all stages of maturity, the highest concentration is in ripe fruits. In some instances, an animal can be poisoned by eating 0.1 to 0.3 percent of its weight in silverleaf nightshade.

Livestock signs

The glycoalkaloid can cause two types of effects. Nervous effects include:

- Incoordination
- Excessive salivation
- Loud, labored breathing
- Trembling
- Progressive weakness or paralysis
- Nasal discharge

Effects of gastrointestinal irritation include:

- Nausea
- Abdominal pain
- Vomiting
- Diarrhea, sometimes with blood

Postmortem examinations in some cases have revealed yellowish discoloration of the body fat.

Plant material may be identified in rumen content of dead animals. In cases of fruit poisoning, many small, tomatolike seeds may be found between the folds of the omasum and in the abomasum.

Integrated management strategies

Veterinarians have had some success administering pilocarpine or physostigmine after the animals were removed from infested pastures. Move affected animals as little as possible and give them good-quality hay and water.

Because silverleaf nightshade is relatively unpalatable, problems usually occur after serious overgrazing or if nightshade is baled up with hay. Do not feed livestock from the ground where many ripe nightshade fruits are available.

If infestations become severe, apply Grazon P+D® at 0.6 to 0.9 pound a.i./acre as an aerial or ground broadcast treatment in the spring when plants begin to flower. For individual plant treatments, mix Grazon P+D® as a 1 percent solution in water.

Mechanical control practices that disturb the soil surface may make the plant infestations more severe.

Fruit ↗

Whole plant ↘

↙ Flower

207

Black Nightshade
Solanum ptycanthum

Black nightshade, also called deadly nightshade, was known in the past as *Solanum americanum* or *Solanum nigrum*. This plant is a dark green, slender-branched, hairless annual growing as tall as 3 feet.

The gently pointed oval leaves have smooth margins. White or purple-tinged flowers are about 0.25 inch in diameter. They give rise to small clusters of round, green fruit that turn black at maturity.

Distribution and habitat

Black nightshade is found across Texas and most of the eastern half of the United States. Within Texas, it is more abundant in the eastern half. These plants often grow in thickets, openings in woods and in disturbed soil and spread into cultivated fields. Regions: 1, 2, 3, 4, 5, 6, 7, 8, 9, 10.

Toxic agent

The toxic agents of black nightshade are glycoalkaloids like those found in the other nightshades. This plant, though called deadly nightshade, is probably less toxic than some of the other common species. This is not the same plant as the European black nightshade (*Solanum nigrum*).

Its broad distribution and frequent evidence of having been grazed without associated livestock poisonings support its relative safety. Unlike many other nightshade species, its fruit is suspected to be more toxic when green than when ripe. Poisonings occur only when large amounts are consumed.

Livestock signs

The signs of poisoning may be related to the nervous system or the gastrointestinal tract.

They may include:
- Anorexia
- Depression
- Excessive salivation
- Diarrhea or constipation
- Muscle trembling
- Weakness
- Colic

The very small seeds may be found in the gastrointestinal contents of dead animals.

Integrated management strategies

Avoidance is the best insurance against this plant. At times, black nightshade may almost cover pens that have been vacant for some time. Do not use the pens if green fruits are present.

↖ Fruit

Whole plant ↗

Flower ↓

Buffalobur
Solanum rostratum

Buffalobur is a prickly annual of the nightshade family typically growing up to 2 feet tall. The leaves, which vary in shape and size, are irregularly rounded and deeply lobed with spiny veins.

The yellow flowers appear from May to October, and the fruit is enclosed by a prickly bur.

Distribution and habitat

Buffalobur is considered a weed nearly everywhere it grows. It is common in old fields, roadsides, overgrazed pastures and disturbed areas and near water tanks throughout Texas.

A native of the Great Plains, it is found growing from North Dakota to Texas, and westward and south into Mexico. Regions: 1, 2, 3, 4, 5, 6, 7, 8, 9, 10.

Toxic agent

Buffalobur can poison horses, sheep, goats and cattle. However, sheep and goats are more resistant than cattle, and in controlled experiments, goats were not poisoned at all. Its toxic agent is the glycoalkaloid solanine.

The leaves and fruit contain solanine at all stages of growth. In some instances, as little as 0.1 to 0.3 percent of an animal's weight in buffalobur is enough to be toxic.

Species within the genus *Solanum* can also accumulate excess nitrates in soils that are high in nitrogen.

Livestock signs

The glycoalkaloid can cause two types of effects in a poisoned animal. Nervous effects include:

• Incoordination
• Excessive salivation
• Loud, labored breathing
• Trembling
• Progressive weakness or paralysis
• Nasal discharge

Effects of gastrointestinal irritation include:
• Nausea
• Abdominal pain
• Vomiting
• Diarrhea, sometimes with blood

Typical nitrate signs may also be exhibited but are much less common. Plant material may be identified in rumen content of dead animals.

Integrated management strategies

Because buffalobur is unpalatable and mechanically injurious to the mouth, problems occur only in unusual circumstances. Some cases have occurred in cattle grazing very lush wheat pastures (no roughage) suddenly gaining access to areas infested with mature, dead buffalobur. Others have resulted from extreme overgrazing.

Good range management practices can reduce the incidence of livestock poisoning.

If infestations become severe, apply Grazon P+D® at 0.6 to 0.9 pound a.i./acre as an aerial or ground broadcast treatment in spring when plants begin to flower.

For individual plant treatments, mix Grazon P+D® as a 1 percent solution with surfactant and water, and thoroughly wet the leaves using a hand-held sprayer.

Mechanical control practices that disturb the soil surface may make the plant infestations more severe.

Flower ↗

Whole plant ↓

Texas Mountain Laurel, Mescal Bean
Sophora secundiflora

Mescal bean is a woody evergreen shrub less than 10 feet tall or, in limited areas, a tree growing up to 35 feet tall. It is a member of the legume family.

The leaves are alternate with seven to 13 leaflets. The flowers are purple and strongly fragrant.

The fruit is a large, hard, woody, jointed, one- to eight-seeded legume pod. The seeds are bright red, hard-surfaced and about 0.5 inch long.

Distribution and habitat

This plant is most common on limestone hillsides and canyons. In Texas, it grows in the southern edge of the Edwards Plateau, in shallow soils of the Rio Grande Plains and occasionally in the western part of the Edwards Plateau and Trans-Pecos regions. It also grows in New Mexico and Mexico. Regions: 2, 5, 6, 7, 10.

Toxic agent

Quinolizidine alkaloids are the toxic agents in mescal bean. Sheep and goats are poisoned under range conditions. Cattle are also susceptible, based on results from feeding experiments. Individual animals vary considerably in susceptibility. Sheep and goats fed toxic doses of about 1 percent of their weight in mature foliage were susceptible to attacks for up to 12 days.

Whole seeds pass through the digestive system intact and harmless. Ground or chewed seeds are more toxic than mature foliage, which is considerably more toxic than young foliage.

Livestock signs

Signs are nervous in character and rarely fatal unless broken or masticated seeds are involved. Signs include:

- Exercise-induced trembling
- Stiff gait
- Falling, with inability to rise
- Returning to normal within 5 minutes if left alone

Forced exercise may cause signs to return.

Integrated management strategies

When other forage is scarce, give animals supplemental feed to prevent them from eating poisonous amounts of mescal bean. Confine affected animals and feed them a good ration until they recover.

Good range and grazing management practices usually eliminate problems with poisoning.

↖ Pods and seeds

Flower and leaf ↗

Whole plant ↓

213

Johnsongrass
Sorghum halepense

Johnsongrass is a vigorous, coarse, perennial grass with scaly root stalks. It reproduces by underground rhizomes and seeds.

This grass has broad leaves and grows 3 to 6 feet tall. The numerous seeds that develop in the fall are yellow to purplish, occurring in a large, spreading, open seed head.

Distribution and habitat

Johnsongrass was introduced into South Carolina from Turkey in the early 1800s. Since then, it has spread across most of the southern United States.

Johnsongrass is considered a weed in cultivated fields. It grows well in disturbed soils, along irrigation ditches and stream bottoms. In some situatons, it is valued as a forage and erosion control plant. Regions: 1, 2, 3, 4, 5, 6, 7, 8, 9, 10.

Toxic agent

Sorghum forages under the stress of rapid growth or drought generate cyanogenic glycosides, which are converted to free cynanide in the rumen. Free cyanide may be present in freshly frosted plants. Most losses from johnsongrass are caused by cyanide poisoning. All domestic animals are susceptible to cyanide; ruminants are the most susceptible.

The plant can also accumulate dangerous levels of nitrates after fertilization and during drought.

In most instances, the plant is a fair forage, but it becomes toxic under conditions favoring cyanide or nitrate accumulation.

Livestock signs

Cyanide is one of the most rapidly acting poisons. Signs of illness may begin within 5 minutes after consumption. Death may occur within 15 minutes or several hours.

Clincal signs generally occur in this order:
- Salivation and labored breathing
- Muscle tremors
- Incoordination
- Bright red venous blood
- Convulsions
- Death

High nitrate levels in the plant can complicate the problem and produce nitrate poisoning in sheep and cattle (see Descriptions of Animal Conditions).

Sorghum cystitis may be produced in horses. This disease is characterized by loss of control of the rear legs and bladder, resulting from permanent damage to the spinal cord. The agent that causes this condition is unknown.

Integrated management strategies

Even plants containing high levels of cyanogenic glycosides might not poison livestock. Because poisoning depends on the presence of free cyanide, anything preventing its development in the digestive system can lessen or eliminate the danger of poisoning. Certain feeds, such as alfalfa hay and linseed cake, retard cyanide release.

Treat livestock showing signs typical of cyanide poisoning with sodium nitrite and sodium thiosulfate. If nitrate poisoning is the problem, treat with methylene blue.

The best preventive measure is to defer infested pastures during danger periods.

Only a small percentage of horses consuming sorghum develop cystitis. However, there is no treatment and the animals must be destroyed. Therefore, never feed sorghum forage or hay as a staple diet to horses.

Whole plant ↗

Seed head →

↙ Whole plant

Queen's Delight, Trecul Stillingia
Stillingia spp.

All three species of *Stillingia* in Texas are hairless, perennial herbs with milky sap and numerous shoots arising from a woody base.

Two of the species are similar, with alternate, long, narrow leaves having a serrated edge and a gland in the notch of each serration. The third species has broader, more rounded leaves. Small, inconspicuous flowers appear in a spike at the end of each stalk.

Distribution and habitat

One or more species grows in each vegetational area of Texas. One species can be found in loose sands, particularly in East and West Texas. The other two grow in calcareous soils of the Rio Grande Plains and the Edwards Plateau. Regions: 1, 2, 3, 4, 5, 6, 7, 8, 9, 10.

Toxic agent

These plants are reported to contain cyanogenic glycosides, which release free cyanide in the rumen. They have a very low palatability and are not consumed by livestock except in extreme circumstances. Only one species, *Stillingia treculiana* (trecul stillingia), is a significant threat and has poisoned sheep in drought conditions.

Livestock signs

Soon after consuming the plant, sheep may display clinical signs including:

• Salivation
• Muscle tremors
• Incoordination
• Bloating
• Bright red venous blood
• Convulsions
• Death from respiratory failure

Death may occur rapidly after consumption of the plant, and most animals poisoned on pastures are found dead before clinical signs are observed.

Integrated management strategies

Good range management practices allowing adequate desirable forage will prevent poisoning by these plants. Sick animals will respond to intravenous administration of sodium nitrite followed by the intravenous administration of sodium thiosulfate.

↑ Queen's delight

Trecul stillingia ↓

217

Dutchman's Breeches
Thamnosma texana

Dutchman's breeches is a perennial weed that is a member of the citrus family. The plant is usually about 6 inches tall with simple, alternate leaves. The flowers vary from canary yellow to purple with a slight yellow center and occur from about May to October.

Dutchman's breeches is most easily identified by its strong, aromatic smell when bruised. Its fruit is unique, shaped like inflated Dutchman's breeches with the legs projecting upward.

Used medicinally by American Indians of the Southwest, the plant has been eaten as a tonic and used to treat gonorrhea.

Distribution and habitat

This plant is distributed widely in gravelly calcareous soils in Texas, New Mexico, Arizona and California. In Texas, it is common in the Rio Grande Plains, southern and western parts of the Edwards Plateau, the Trans-Pecos and southern parts of the Plains. Regions: 2, 6, 7, 8, 9, 10.

Toxic agent

The toxic properties of Dutchman's breeches can be attributed to the presence of psoralens (strong photosensitizing agents) in the plant.

Sheep consuming about 1 percent of their body weight of this plant have shown signs of primary photosensitization within 24 to 48 hours. Cattle and goats also have shown primary photosensitization when grazing areas with large amounts of Dutchman's breeches. Humans who get plant sap on their skin suffer severe sunburn in contacted areas.

Livestock signs

Signs of primary photosensitization include:

- Increased body temperature
- Photophobia (animals avoid sunlight)
- Edema of the muzzle, ears and vulva
- Inflamation and edema of the eye

Integrated management strategies

Sheep and goats like the taste of Dutchman's breeches and can be expected to graze this weed, perhaps heavily in areas where more palatable forage plants are less available.

Most poisoning cases have occurred when sheep or goats were placed in pastures that for several years had been grazed only by cattle. Where sheep and goats have historically grazed continuously, the plant has been almost eliminated and no longer threatens them. However, the plant may still be abundant over the fence along a highway.

Pasture rotation and/or supplemental feeding may help prevent consumption of Dutchman's breeches.

When the first cases of photosensitization occur, move the animals to new pasture if possible. Place sick animals in the shade with feed and water. Painting or spraying the affected skin areas with methylene blue solution or some other nontoxic dye is beneficial. Avoid applying grease, oils or tannic acid.

Pod and flower ↗

Whole plant ↓

Goathead,
Puncturevine
Tribulus terrestris

Goathead is an annual weed in the caltrop family. The prostrate stems radiate from a tap root and bear pairs of opposite leaves.

The flattened fruit resembles a goat's head. It breaks into five nutlets, each bearing two strong, woody spines, hence the name puncturevine. The flowers are small and have five yellow petals.

Distribution and habitat

Goathead is an introduced weed from Europe. Widely distributed in disturbed areas and along trails and roadsides, it may abound in severely overgrazed pastures. It is found throughout Texas except on the Gulf Coast and extreme eastern part of the state. Regions: 2, 3, 4, 5, 6, 7, 8, 9, 10.

Toxic agent

The plant causes hepatogenous photosensitization in sheep and possibly also in cattle. All parts of the plant are toxic at all growth stages, but wilted plants are the most hazardous.

Goathead also can accumulate high levels of nitrate. The spiny burs this plant produces are mechanically dangerous, producing lesions on the mouth or feet.

Livestock signs

In natural cases, typical lesions of severe hepatogenous photosensitization were seen, including:

• Blindness

• Peeling of light-colored skin

• Loss of lips and ears

• High mortality of young animals

Nitrate poisoning signs also may be evident (see Description of Animal Conditions).

Integrated management strategies

The best way to reduce potential livestock losses from goathead is to adopt good range management practices. Given the opportunity, animals avoid this plant, preferring more palatable forage species. If livestock are eating goathead, chances are that stocking rates are too high.

Remove livestock showing signs from infested pastures and provide shade, a good quality diet and water.

Most broadleaf herbicides control the plant easily, but use caution when treating with 2,4-D, as this chemical increases nitrate accumulation in the plant.

Mechanical improvement techniques that disturb the soil surface may increase infestation of this plant for a short time after treatment.

↖ Leaf

Flower
and seed ↗

Whole plant ↓

221

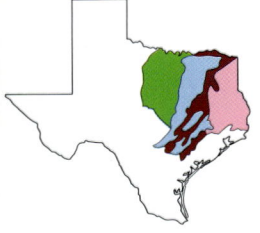

Hairy Vetch
Vicia villosa

Hairy vetch is an annual or biennial spreading herb with climbing stems growing up to 3 feet long. The leaves have long, soft hairs with 10 to 20 leaflets borne opposite each other and tendrils at the end.

The flowers, borne in a dense one-sided spike, are violet and white to rose colored. The oblong pod has a beak-shaped tip, grows up to 1.25 inches long and opens at maturity.

Distribution and habitat

Hairy vetch has been widely introduced and cultivated with cereal grains as a cool-season forage in the eastern third of Texas. It has escaped in many areas, and dense stands are found along roadsides and in pastures where it has been allowed to seed.

A native species, *Vicia leavenworthii*, which extends west to Schleicher County, is associated with cases of vetch poisoning in rare instances where there was adequate fall and winter rainfall for abundant growth. Regions: 1, 3, 4, 5.

Toxic agent

Neither the toxic agent nor the specific conditions in which poisoning develops are known.

Poisoning invariably involves animals with black pigmented skin (Angus, Angus cross or Holstein cattle; black horses).

Poisoning usually occurs after the plants begin to flower, but some cases have occurred in Central Texas when the plants were in the seedling stage with runners no more than 8 inches long. Years of uneventful grazing of vetch can pass between poisoning episodes.

Livestock signs

The first clinical signs of affected animals appear similar to photosensitization, except that the lesions are on black skin. Clinical signs may include:

• Inflamed skin (head, neck and tailhead)

• Thickening of skin with granular tumors

• Matting of hair and sloughing of skin

• Diarrhea

• Anorexia

• Loss of weight

• Death

The death rate of affected animals is high. They often die from kidney failure, as granulomas grow in the various internal organs, including the kidneys.

Integrated management strategies

Hairy vetch is a desirable, nutritious forage and should continue to be used as such. However, cattle grazing vetch should be observed frequently. Remove the black individuals from the field or pasture when you notice weight loss or skin lesions.

Flower ↗

Pod ↘

↙ Whole plant

Cocklebur
Xanthium spp.

Texas has six species of cocklebur. They are coarse, rough, annual weeds with alternate, toothed or lobed leaves.

Separate male and female flowers grow on the same plant, although both are inconspicuous. Male flowers occur in dense clusters on the ends of the stems; female flowers occur in leaf axils.

Many spines cover the conspicuous cocklebur fruits, which have two compartments, each with a seed.

Distribution and habitat

Cockleburs are found throughout most of the United States. In dry areas they are most common around water holes, playas, arroyos and disturbed areas. Regions: 1, 2, 3, 4, 5, 6, 7, 8, 9, 10.

Toxic agent

Cocklebur poisons all classes of livestock. The toxic substance in the seeds is carboxyatractyloside, a glycoside causing hypoglycemia and massive liver damage.

Although livestock generally do not eat the seeds, problems can occur when cattle are fed whole cottonseed or hay contaminated with cocklebur. The toxic agent remains present in the seedling through the cotyledon stage. The toxin concentration drops rapidly when the first true leaves appear.

A toxic dose of seedlings is about 0.75 to 1.5 percent of the animal's weight. Seedlings are toxic even when dead and dry.

Livestock signs

Signs generally occur 12 to 48 hours after cocklebur seedlings are eaten. They include:

- General weakness
- Depression
- Unsteady gait
- Rapid, labored breathing with a weak, rapid pulse
- Subnormal body temperature with nausea and regurgitation

Once the animal is down, it convulses, makes running motions with its legs or shows a marked curvature of the neck. Death usually occurs a few hours to 3 days after the first signs appear.

Integrated management strategies

Keep poisoned animals warm and give them large amounts of fatty substances. Cream, milk and mineral oil can be given by mouth, administered through a stomach tube to avoid producing inhalation pneumonia. Heart and respiratory stimulants are also recommended.

Poisoning usually occurs when many seedlings germinate after a rain or as water recedes in low-lying areas. Hazard is greatly increased when seedlings are commingled with desireable foliage. Do not feed supplements or hay on ground where seeds or seedlings are present.

Control cockleburs by aerial or ground broadcast application of 2,4-D at 1.0 pound a.i./acre before plants flower. Or, mix a 1 percent solution of 2,4-D for individual plant treatments.

Cocklebur germinates after summer rains, so chemical application may be required more than once a year. Follow any chemical treatment with proper range and livestock management programs.

Any means of mechanical control that prevents the plant from producing seeds helps reduce the population of this annual plant. Concentrations of small seedlings can be burned with a hand-held propane burner.

X. spinosum whole plant ↗

↙ ↓ Whole plant and seedling

Nuttall Deathcamas, Deathcamas

Zigadenus nuttallii

Nuttall deathcamas is a perennial herb arising from a bulb with a black, papery outer coating. Its unbranched, erect, leafy stalk grows to 15 to 30 inches tall.

The mostly basal, curved leaves may be up to 15 inches long on larger specimens. The stalk terminates in a yellowish-white spike of flowers that give rise to egg-like seed capsules.

Distribution and habitat

Nuttall deathcamas is found in the eastern third of Texas on open prairies, on hillsides with calcareous rocks and in post oak areas. This plant is seldom noticed except for the short period when it is in bloom. Regions: 1, 3, 4, 5, 7.

Toxic agent

Deathcamas contains alkaloids toxic to all livestock species, but it causes very few poisonings because it is unpalatable. Animals consuming as little as 0.25 percent of their body weight of green plants may display signs of poisoning in a few hours.

Sheep have been known to eat the young plants in early spring when other forage is scarce. However, most of the deathcamas in Texas grows where there are few sheep. Humans have been poisoned after mistaking the bulbs for onions.

Livestock signs

The clinical signs may include:
- Salivation
- Vomiting
- Depression
- Weakness
- Weak irregular pulse
- Difficulty breathing
- Coma
- Death

Plant material may be identified in the rumen of dead animals.

Integrated management strategies

These plants are conspicuous when in flower and usually occur in small plant communities in a pasture. Mechanical removal is recommended, but care must be taken to dig out the bulb, which may be a foot or more below the surface.

In early spring, do not allow sheep to graze pastures with large populations of deathcamas without adequate supplemental feed.

← **Whole plant**

Flower ↗

Bulb ↘

227

Abomasum: The fourth compartment of the ruminant stomach. Called the "true stomach" because this is the compartment where acid digestion occurs.

Acute: A rapid, severe onset of signs.

A.I. (a.i.): Active ingredient in a pesticide formulation.

Alkaloids: A very broad category of complex organic bases containing nitrogen and usually oxygen. These often have a bitter taste.

Alternate: Leaf arrangement with one leaf per node, at different heights on the stem.

Anorexia: Loss of appetite.

Awn: Slender bristle, usually at the end of a plant organ.

Axial: Associated with branches instead of the main stem.

Axil: Angle formed by the stem and an appendage (e.g., leaf).

Basal: Associated with the base of the plant.

Bract: Modified leaf associated with a flower.

Calystegine: Class of alkaloids that inhibit glycosidases, allowing buildup of certain cellular products, resulting in dysfunction. These affect many cell types, but can permanently damage nervous tissues (brain, spinal cord).

Calyx (calyces): The outermost organs of a flower; usually green.

Capsule: Dry, multiseeded fruit, usually opening at maturity.

Cardiac glycoside: Sugar-containing compound that affects the heart.

Cholestasis: Accumulation of bile in the liver.

Chronic: Over a period of time. Can refer to prolonged or repeated exposure to toxins or to the progression of clinical signs.

Cirrhosis: Liver disease involving hepatocellular necrosis, replacement of dead tissue with fibrous tissue (fibrosis), and distortion of microscopic structural units and, therefore, their function. Usually affects the organ as a whole (diffuse).

Cleft: Deeply indented; split.

Compound: A leaf structure that is divided into distinct segments (leaflets), which are pinnately (like a feather) or palmately (like an open hand) arranged.

Coumarin: A specific two-ringed organic chemical structure containing oxygen.

Cyanogenic glycoside: A sugar-containing compound that also contains a releasable cyanide group.

Diterpene alkaloid: A 20-carbon compound structure including nitrogen; not water-soluble. Often part of the essential oil of the plant. Terpenes are plant equivalents of steroids.

Edema: Accumulation of fluid within tissues, usually resulting in swelling.

Endophyte: One organism that lives within another. In these instances, a fungus that lives within its plant host, inseparable.

Euthanasia: An induced, easy or painless death.

Forb: Any nonwoody plant (herb) other than a grass.

Furocoumarin: Psoralen. A multi-ringed organic chemical that increases the light sensitivity (primary photosensitization) of the skin.

Gastroenteritis: Inflammation of the lining of the gastrointestinal tract, often accompanied by outward signs of abdominal pain and diarrhea, sometimes with evidence of blood.

Glycoalkaloid: Sugar-containing alkaloid.

Glycoside: Sugar-containing chemical compound.

Gypsiferous: High in sulfate, as a soil, or water.

Hepatic: Relating to the liver.

Hepatotoxic: Toxic to the liver.

Hilum (hilus): Scar on the seed at its point of attachment.

Hypocalcemia: Abnormally low blood calcium.

Hypoglycemia: Abnormally low blood sugar.

Jaundice: Bile accumulation in the blood, which results in an abnormal yellow color of the skin, mucous membranes, eyeballs and/or fat.

Jugular distension: Bulging of the major veins in the neck, a sign of excessive blood pressure.

Leaflet: One segment of a compound leaf.

Lectin: Toxalbumin; a plant protein with attached sugars (glycoprotein) that interferes with basic cellular function throughout the body.

Legume: 1. Member of family Leguminosidae. 2. Seed pod typically produced by members of that family.

Lesion: Injured or abnormal area of tissue.

Lethal dose: The amount of a toxin that causes death (as compared to a toxic dose, which causes only the appearance of signs of poisoning).

Mesentery: Membranous fold attaching various organs to the body wall.

Monensin: Ionophore antibiotic that changes the types of microorganisms in the rumen to improve feed efficiency; also selects against thiaminase producing microbes, helping prevent polioencephalomalacia.

Narcotic: Substance producing insensibility or stupor.

Necrosis: Tissue death.

Neurotoxic: Poisonous or destructive to nerve tissue.

Omasum: Small fermentation compartment of the ruminant stomach between the rumen and abomasum.

Opposite: Leaves arranged two at each node in opposite directions.

Palatable: Attractive to eat; tasty.

Panicle: Arrangement of flowers around a central axis with individual stalks supporting each flower.

Petiole: Leaf stalk, attaching it to the stem.

Phorbol ester: Certain plant compounds with complex structures associated with cancer promotion.

Photodynamic: Activated by light.

Photosensitization: Hypersensitivity to light; results in photophobia (shying away from light) and sunburn on light-colored and lightly haired skin; sunburn can progress to the point of sloughing skin. Frequently referred to as "photo." See Animal Conditions section for more information.

Photosensitization, hepatogenous: Hypersensitivity to light caused by an agent that damages the liver, impairing its ability to metabolize chlorophyll. A partial breakdown product of chlorophyll (phylloerythrin) then causes photosensitization.

Photosensitization, primary: Hypersensitivity to light caused directly by a toxin itself.

Phytate: Plant acid compound.

Polioencephalomalacia: Degeneration of the grey matter of the brain. Most often associated with sudden dietary changes that shift the rumen microbial population to include a microbe that destroys thiamine. Also produced by ingestion of certain plants that contain a thiaminase.

Prostrate: Lying flat upon the ground. Applies both to plant growth habit and to an animal in a state of absolute exhaustion or weakness (prostration).

Pulmonary adenomatosis: A condition of the lungs in which the tissue becomes thickened and gas exchange can no longer take place. Literally, "glandularization of the lungs." The lung tissue fills with air that cannot escape.

Purgative: Causing evacuation of the bowels.

Pyridine alkaloid: Single-ring compounds containing nitrogen.

Pyrrolizidine alkaloid: Organic structure based on five-membered rings, including nitrogen, which causes progressive liver cirrhosis.

Quinolizidine alkaloid: Organic structure based on six-membered rings including nitrogen causing a variety of effects depending on attached functional groups. Several cause birth defects.

Resinoid: One of a wide variety of chemical compounds classified by their physical characteristics on purification. These are solid or semisolid at room temperature and are not water-soluble—otherwise they may have nothing in common.

Reticulum: Small compartment in the ruminant stomach closely associated with the rumen and with roughage fermentation. The papillae transform into ridges on the interior surface to form a "honeycomb" configuration.

Rhizomes: Underground modified stems producing leafy shoots above and roots below.

Rosette: Crowded cluster of leaves at ground level radiating from a central root.

Rumen: By far the largest digestive compartment in the ruminant stomach. Its primary function is to house microorganisms for fermentation of roughage. The interior surface is densely covered with papillae (fingerlike projections) that increase its surface area for nutrient absorption.

Rumen stasis: Lack of rumen activity. Usually involves death or inactivity or rumen microbes; also, loss of motility of the rumen walls.

Saponin: Sugar-containing sterols that irritate the gastrointestinal tract. When absorbed into the bloodstream, they can cause red blood cell rupture and liver damage.

Serous fluid: Clear body fluids resembling the clear, watery portion of the blood.

Serrate: Saw-toothed; with teeth pointing forward.

Sesquiterpene lactone: Organic compound with a steroidal backbone, nitrogen and oxygen. Usually, dermal and gastrointestinal irritants. Sometimes, allergic sensitizers.

Sessile: Attached directly at the base, with no stalk.

Spike: Flowers arranged around a central axis, with each flower on its own stem. Same as panicle, above.

Stamen: The male organ of the flower, which bears the pollen.

Stocking up: In horses, swelling of the lower legs (peripheral dependent edema).

Terminal: Growing at the end of a branch or stem.

Toxic dose: The amount of a toxin needed to induce signs (not necessarily a lethal dose, which causes death).

Tremorgen: Fungal byproduct (mycotoxin) that causes trembling because of neurotoxicity.

Triterpene: Plant version of a steroidal compound. Various effects depending on attached functional groups. Part of the plant's essential oil.

Tropane alkaloid: Nitrogen-containing organic ring structure, akin to atropine.

Trypsin inhibitor: Agent that inhibits the digestive enzyme, trypsin. By association, these are expected to inhibit other enzymes as well, disrupting cellular functions.

Winged stem: Stem with a pair of parallel, thin extensions on both sides. Usually the wings are extensions of clasping leaves down the stem.

Cheeke, Peter R., *Natural Toxicants in Feeds, Forages and Poisonous Plants*, 2nd ed. Danville, IL: Interstate Publishers, Inc., 1998.

Colegate, Steven M. and Peter R. Dorling, eds., *Plant-Associated Toxins: Agricultural, Phytochemical and Ecological Aspects*. Wallingford, Oxon, UK: CAB International, 1994.

Correll, D.S. and M.C. Johnston, *Manual of the Vascular Plants of Texas*. Renner, TX: Texas Research Foundation, 1970.

Garland, Tam and A. Catherine Barr, eds., *Toxic Plants and Other Natural Toxicants*. Wallingford, Oxon, UK: CAB International, 1998.

Hatch, Stephan L., Kancheepuram N. Gandhi and Larry E. Brown, *Checklist of the Vascular Plants of Texas*. Texas Agricultural Experiment Station, 1990.

Keeler, Richard F., Kent R. Van Kampen and Lynn F. James, eds., *Effects of Poisonous Plants on Livestock*. New York: Academic Press, 1978.

Kingsbury, John M., *Poisonous Plants of the United States and Canada*. Englewoods Cliffs, NJ: Prentice-Hall, Inc., 1964.

McGinty, Allan, J.F. Cadenhead, Wayne Hamilton, Wayne C. Hanselka, Darrell N. Ueckert and Steven G. Whisenant, B-1466, *Chemical Weed and Brush Control Suggestions for Rangelands*. Texas Agricultural Extension Service, 2000.

Sperry, O.E., J.W. Dollahite, G.O. Hoffman and B.J. Camp, B-1028, *Texas Plants Poisonous to Livestock*. Texas Agricultural Extension Service, 1968.

Index

Index to Common and Scientific Plant Names

236

Index

239

241

Index

242

243

Field Key

	Abnormal leaflower	African rue	Beef-steak plant	Berlandier lobelia	Berlander nettle spurge	Bermudagrass	Bishop's-weed	Bitter sneezeweed	Black locust	Bracken fern	Broom snakeweed	Buttercup	Buttonbush	Carelessweed	Castor-bean	Chinaberry	Chinese tallow	Cocklebur	Conzya	Coyotillo
Cattle	c	c	c	c		c	c	c	c	c	c	c	c	c	c		c	c	c	c
Sheep	s	s	s		s	s	s		s	s	s		s	s	s			s	s	s
Goats	g				g	g	g		g	g	g		g	g	g			g	g	g
Horses		h	h		h	h		h	h	h	h		h	h	h		h			
Pigs															p		p			

Livestock signs

	Abnormal leaflower	African rue	Beef-steak plant	Berlandier lobelia	Berlander nettle spurge	Bermudagrass	Bishop's-weed	Bitter sneezeweed	Black locust	Bracken fern	Broom snakeweed	Buttercup	Buttonbush	Carelessweed	Castor-bean	Chinaberry	Chinese tallow	Cocklebur	Conzya	Coyotillo
Abdominal pain									✓		✓	✓								
Abnormal heartbeat								✓								✓				
Abnormal urination		✓							✓	✓										
Abortion											★		✓							
Anorexia	✓	✓						✓	✓	✓				✓	✓	✓				
Arched back																				
Birth defects																				
Blindness																	✓			
Bright red blood																				
Colic											✓			✓	✓					
Collapse				✓																✓
Coma											✓					✓				
Continous walking	✓																	✓		
Convulsions											✓			✓	✓					
Delayed signs																				
Depression/weakness	✓	✓		★	✓		✓	✓		✓	✓		✓	✓	✓	✓	✓		✓	
Diarrhea	✓			★			✓	✓		✓	★			✓	✓	✓				
Dilated pupils											✓									
Excess salivation		✓					✓			✓										
Excitability	✓			✓		✓					✓							✓	✓	
Green salivation																				
Incoordination		✓			✓		✓	✓		✓			✓		✓		✓		✓	
Irregular breathing			★		✓								✓		✓					
Loss of weight								✓	✓											✓
Nitrate poisoning											★									
Photosensitization	★					✓	★													
Pushing on objects																				
Retained placenta											✓									
Running into objects																				
Stiffness		✓						✓				✓								
Sudden death											✓			✓		✓				
Trembling		✓			✓			✓		✓			✓				✓			
Unable to eat/drink				✓																
Unable to rise				✓		✓														✓
Unthrifty offspring									✓											
Vomiting/regurgitation								✓			✓			✓		✓		✓		✓

*This table is intended as a diagnostic tool for field use. Further consultation on individual plants and specific poisoning syndromes, discussed in more detail in the text, is suggested.

★ Indicates the most common signs.

Dallisgrass ergot	Desert baileya	Desert spike	Desert tobacco	Dock	Dutchman's breeches	Goathead	Golden corydalis	Gordon bladderpod	Groundsels	Guajillo	Hairy caltrop	Hairy vetch	Hog-plum	Inkweed	Jimmyfern	Jimsonweed	Johnsongrass	Kleingrass	Knotweed	Kochia	Lanceleaf salvia	Lantana	Larkspur	Leatherstem	Lechugilla	Locoweeds
c	s	c	c	c	c	c	c		c	c	c	c		c	c	c	c	c	c	c	c	c	c		c	c
s		s	s	s	s	s	s		s	s	s		s	s	s	s	s	s	s	s	s	s	s		s	s
	g		g	g	g				g	g	g			g	g	g	g	g	g		g	g	g		g	g
h		h	h	h		h	h		h			h				h	h	h	h	h	h	h	h		h	h

All signs may not appear in all cases and some may be seen only in one species.
Number of signs may or may not be indicative of relative toxicity.
"Delayed signs" are those that may appear 1 to several months after ingestion of poisonous plants.

Field Key

	Maples	Mesquite	Milkweeds	Mountain laurel	Mountain mahogany	Mountain pink	Narrowleaf sumpweed	Nightshades	Nuttall deathcamas	Oak	Oleander	Paperflowers	Peavine	Poison hemlock	Pokeberry	Queen's delight	Rainlily	Rayless goldenrod	Red buckeye
Cattle		c	c	c	c	c	c	c	c	c	c	c	c	c	c	c		c	c
Sheep		s	s	s	s	s		s	s	s	s	s	s	s	s	s	s	s	s
Goats		g	g	g	g	g		g	g	g	g	g	g		g	g	g	g	
Horses	h	h	h		h			h	h	h	h			h	h	h		h	h
Pigs														p		p			
Livestock signs																			
Abdominal pain		✓		✓			✓		✓	✓			✓	✓					
Abnormal heartbeat			✓					✓		✓				✓					
Abnormal urination	✓							✓	✓									✓	
Abortion					★														
Anorexia	✓	✓		✓				✓		✓	✓								
Arched back																		✓	
Birth defects														✓					
Blindness																	✓		
Bright red blood			✓											✓					
Colic		✓					✓	✓	✓	✓			✓	✓					
Collapse			✓					✓											
Coma		✓						✓	✓		✓			✓			✓		
Continous walking																			
Convulsions		✓		✓											✓				
Delayed signs																			
Depression/weakness	✓		✓					✓	✓	✓	✓	✓	✓	✓				✓	✓
Diarrhea				✓				✓							✓				
Dilated pupils			✓																
Excess salivation	✓	✓		✓				✓	✓		✓			✓	✓				
Excitability			✓											✓					
Green salivation										★									
Incoordination			✓	✓			★				✓	★	✓		✓			✓	
Irregular breathing			✓					✓	✓					✓	✓	✓	★		
Loss of weight											✓	✓							
Nitrate poisoning								✓											
Photosensitization																	★		
Pushing on objects																			
Retained placenta																			
Running into objects																			
Stiffness			✓											✓			✓		
Sudden death		★							★						★				
Trembling		✓		✓	✓		★							✓	✓			★	★
Unable to eat/drink		★																	
Unable to rise								✓											
Unthrifty offspring																✓			
Vomiting/regurgitation								✓	✓		✓			✓		✓	✓		

*This table is intended as a diagnostic tool for field use. Further consultation on individual plants and specific poisoning syndromes, discussed in more detail in the text, is suggested.

★ Indicates the most common signs.

Russian thistle	Ryegrass ergot	Sacahuista	Sennas	Sesbanias	Silktree, mimosa	Singletary pea	Smallhead sneezeweed	Snow-on-the-mountain	Spotted water hemlock	Starthistles	Sweetclover	Tansy mustard	Tarbush	Texas persimmon	Threadleaf sartwellia	Tobosagrass ergot	Tree tobacco	Western bitterweed	White snakeroot	Whitebrush	Wild indigo	Wild onion	Wild plum	Wright buckwheat
c	c	c	c	c	c	c	c	c		c	c	c	c	c		c	c		c	c		c	c	c
s	s	s	s	s	s	s	s		s		s		s			s	s	s	s	s		s		s
g	g	g	g	g		g	g		g		g		g	g		g		g				g		g
h	h		h	h		h	h	h	h							h		h	h	h	h	h		h
				p																				

All signs may not appear in all cases and some may be seen only in one species.

Number of signs may or may not be indicative of relative toxicity.

"Delayed signs" are those that may appear 1 to several months after ingestion of poisonous plants.